Chavan's Theory of Light

Ripple Alignment and the ATT Principle

A Foundational Report Under Universal Energy Dynamics (UED)

By

Sandeep Chavan

Independent Researcher, Founder of UED

CHAVAN'S THEORY OF LIGHT

First edition. May 11, 2025.

ISBN: 979-8231268689

Written by Sandeep Chavan.

Table of Contents

To all the **scientists, seekers, and silent thinkers**—who dared to question the known and gave us the courage to ask again.

To my **father**, whose strength shaped my vision,and to my **family**, whose unwavering supportcarried me through the ripple of every idea.

This book is for you.And for the light that never needed to move to be seen.

— *Sandeep J. Chavan*

"Light was never a traveler.It was a ripple—waiting for the universeto be ready to see."

— Sandeep J. Chavan

Opening Note

To the Seekers of Truth—Within and Beyond Equations

This report is not a rebellion against science. It is a restoration of its original spirit.

For centuries, we've measured light, traced its rays, and modeled its behavior through lenses and laws. We built magnificent theories—some elegant, some convoluted—but all bound by the same silent assumption: That light flows. That it travels. That it carries something from "there" to "here."

But what if that assumption is the very illusion that held us back?

What if light was never a traveler—but a consequence? Not a wave or a particle, but a ripple alignment—a silent handshake between structured energy and conscious awareness? What if we've been chasing light, while all along, light was waiting for us to notice how we *perceive*?

This document introduces a fundamental shift in our understanding of light—not by adding complexity, but by removing projection. It proposes that the so-called "speed of light" is not a limit, but a misinterpreted metric—replaced in this new view by ATT (Alignment Threshold Time), the minimum ripple-alignment interval necessary for perception to trigger.

We will revisit legacy experiments not to dismiss them, but to decode what they were really detecting. We will abandon imaginary wavefronts and replace them with real ripple structures. We will cross the boundary between optics and metaphysics—not to romanticize science, but to reunite it with consequence-based reality.

To the scientists: Let this be your second lens. To the thinkers: Let this offer clarity where abstraction blurred the view. To the independent researchers: Let this remind you that truth is not patented. And to every curious soul: Let this light not shine *on* things, but *within* you.

Welcome to a new theory. One that does not carry light forward—But helps you stand still enough... to *feel* it.

Preface: Why Rethink Light Now?

For centuries, light has been one of science's most fascinating enigmas—and perhaps one of its most misunderstood. From Newton's corpuscular theory to Maxwell's wave equations, from Einstein's photons to quantum electrodynamics, light has been dissected, modeled, and mathematized in extraordinary ways. We've built devices to measure its speed, used it to define the meter, and even declared its velocity an absolute cosmic limit.

But beneath all this progress lies a deeply ingrained illusion: That light flows.

We have assumed—without ever truly proving—that light behaves like a substance, a beam, or a traveling entity that moves from source to observer. We designed our experiments based on this assumption. We created equations to fit this behavior. And when inconsistencies arose, we patched them with more abstraction, more paradoxes, and more probabilistic uncertainties.

This report does not seek to discredit the work of the past. Rather, it seeks to free us from the assumptions that now limit our vision.

The Illusion of Flow

The idea that light "moves" through space implies a medium, a carrier, or a mechanism of transit. When the aether was disproved, we replaced it with empty space and mathematical formalism. But what if we misunderstood the phenomenon entirely? What if the observer does not *receive* light, but rather *aligns* with it?

In reality, what we call the "speed of light" may not be a true speed at all—but the time it takes for field-level ripple alignment to reach a threshold for perception.

The notion of flow falls apart when we realize that the structured vacuum is not a passive void, but an active field of ripple tension. In this field, "light" is not a thing moving from A to B. It is a signature ripple alignment event—a consequence, not a courier.

Rise of UED: From a Classroom Question to a Cosmic Realignment

Universal Energy Dynamics (UED) was not born in a particle accelerator or in a physics lab. It began in a classroom, with a question as simple as it was profound: *"What is force?"*

From that spark grew a ripple. That ripple has now evolved into a full scientific framework—one that redefines energy, matter, space, time, and now... light.

UED does not reject physics—it refines it by removing projection and restoring consequence. It offers a more elegant, coherent, and meaningful explanation of natural behavior by recognizing that energy is not transmitted—it aligns.

This report builds upon that principle and introduces ATT (Alignment Threshold Time) as the natural replacement for the flawed concept of light-speed.

What This Report Offers

This is not merely a new theory of light—it is a departure from the old paradigm.

- A clean break from the wave-particle duality illusion

- A shift from trajectory to resonance
- A framework that repositions perception, consequence, and energy within a single, unified field theory
- And most importantly, a new way of experiencing light—not as a quantity to be measured, but as a signature to be understood.

Let us then return to light—not as it travels, but as it awakens. Let us see not what it shows, but what it *is*.

Executive Summary

This report presents a foundational rethinking of the nature of light and perception through the lens of **Universal Energy Dynamics (UED)**—a ripple-based framework that replaces classical and quantum optics with a model built entirely on **consequence alignment** in a structured vacuum.

At its core, the report introduces **Alignment Threshold Time (ATT)** as the true determinant of perceptibility, eliminating the need for particle-wave duality, photon travel, and velocity-based interpretations of light. In this model, **light does not move**—it is a ripple pattern that **aligns** with receptive fields. What we perceive as light-speed is simply the **minimum time the field requires to recognize and resolve ripple tension**.

Key Insights and Contributions

- **Reinterpretation of Light**: Light is no longer treated as a flow of particles or wavefronts, but as a **ripple identity seeking resolution** through alignment with structured fields.
- **ATT as the New Metric**: The traditional concept of speed is replaced by **ATT**, a delay metric that reflects how long it takes for ripple-induced tension to trigger a detectable or perceptual event.
- **Collapse of Wave and Ray Optics**: Classical models are shown to be emergent approximations of ripple consequence behavior, with lenses, mirrors, and optical paths functioning as **field-shaping devices**, not beam-directing elements.
- **New Explanations for Optical Phenomena**: Reflection,

refraction, diffraction, and interference are reinterpreted through ripple field interactions—without invoking imaginary wavefronts or discrete particle motion.

- **Experimental Proposal**: A new class of ATT-based detectors is proposed to measure ripple alignment timing, opening a path for empirical validation of ripple optics.
- **Technological Implications**: The report outlines potential applications in **AI vision systems, ripple-based computing, field-aware sensors,** and **ripple-aligned optics** that may transform multiple industries.
- **Philosophical Realignment**: The illusion of light's "speed" is traced back to ATT ≈ 0 in human perception, revealing that our deepest constants may in fact be context-bound consequences of alignment, not fundamental truths.

Conclusion

This report does not offer a modification of existing theories—it offers a replacement. It proposes that **light is not a traveler**, but a ripple awaiting readiness. That **perception is not reception**, but structured consequence. And that the universe responds not with motion, but with **resonance**.

Universal Energy Dynamics introduces not just a new science of light, but a new science of **how reality reveals itself**.

Part I: Deconstructing the Legacy

1. The Myth of Light as a Flow

For centuries, the concept of light has been wrapped in metaphors of motion. Whether described as rays shooting through the air or waves undulating across the void, light has consistently been assumed to be something that *moves*. This assumption, so embedded in our scientific consciousness, was rarely questioned—not even when experiments like Michelson and Morley's seemed to contradict the medium through which such motion might occur.

Albert Michelson's groundbreaking experiments in the late 19th century, especially his rotating mirror method for measuring the speed of light, were deeply rooted in the belief that light travels from a source to a detector. His setup was ingenious, but it rested on a fundamental assumption: that there is a measurable *time-of-flight* for light. In other words, he treated light as if it were a projectile or pulse traveling across a distance, and what he recorded was interpreted as the time taken by this entity to move from point A to point B.

This perspective was carried forward and refined by Einstein, who accepted the constancy of the speed of light and built the special theory of relativity around it. But even Einstein never fully questioned the idea that light *flows*. He accepted the measured speed as a fundamental limit, and the mathematics followed.

However, as elegant as this formulation is, it harbors a hidden presumption: that light is a substance with motion. Measuring the time it takes for light to traverse a known distance assumes that something *moves through space*. This is the invisible assumption that underpins nearly all of modern optics and quantum electrodynamics. And it's this assumption that breaks down when we consider the nature of structured vacuum as introduced in Universal Energy Dynamics (UED).

Michelson's Framework: Measuring What Was Assumed

In Michelson's original apparatus, a beam of light is split, bounced off mirrors, and recombined to detect interference caused by differences in travel time. The underlying idea was simple: if Earth is moving through a medium (aether), then light traveling with and against that motion should show different travel times, just as a boat crossing a river diagonally experiences different speeds depending on current.

But the results were null—no difference in speed, regardless of direction. This famously led to the rejection of the aether. Yet instead of questioning the model of light as a traveling entity, the interpretation merely adjusted: light has no medium, but it still travels. The belief in motion remained untouched.

The apparatus measured *delay*, interpreted as time-of-travel, but what if it instead captured something deeper: the time required for ripple alignment within the structured vacuum between source and mirror?

The Invisible Assumption: Time-of-Travel Implies Substance

Let's be clear: whenever we measure the speed of anything, we implicitly assume it is a *thing*—an object, particle, or entity with definable location and trajectory. Time-of-travel implies that this object exists independently of both emitter and observer and that it makes a journey across space.

But in the case of light, the very act of "receiving" it—perceiving illumination—is not about catching an object. It is about being triggered by a ripple-pattern match.

Imagine striking a bell in a pond. The ripples extend outward not as something leaving the bell, but as a consequence pattern in the water. When another object in the pond responds, it's not because something hit it directly. It's because the structured field—the pond itself—aligned enough to trigger a reaction.

This is the essence of UED. Light is not sent. It is summoned into perceptibility by the field's local alignment.

Therefore, the assumption that light *travels* between source and receiver is an illusion born out of a deeper misunderstanding: that we are separate from the ripple field. In truth, both source and observer are embedded within the same structured vacuum. Perception is not the endpoint of travel—it is the threshold of alignment.

Why This Assumption Breaks Under Structured Vacuum Logic

In UED, space is not empty. It is not a passive stage. It is a structured vacuum field, rich with ripple tensions, interference zones, and equilibrium-seeking behavior. Every energy event introduces ripples—tensions that seek harmony.

What we perceive as light is not a transit, but a ripple alignment event. The source generates a disturbance in the field. That disturbance propagates as a field ripple, carrying not mass or momentum, but pattern identity and energy potential. When the observer's local field reaches resonance with this ripple signature, a perception event occurs.

This event is not caused by something "arriving." It is caused by something "matching."

Here, the concept of speed becomes secondary, even obsolete. What matters is not *how fast* something moves, but *how quickly* resonance is achieved. This leads directly to the concept of Alignment Threshold Time (ATT)—a consequence-based metric far more natural than arbitrary speed.

Michelson's measurements, then, did not record a time-of-travel. They recorded the minimum interval required for field-level ripple alignment under given structural conditions. What he thought was "speed" was really ripple compliance time.

Under this framework, so-called "light speed" becomes a boundary condition—not a velocity. It's the fastest known alignment rate in an ideal, undisturbed structured vacuum. Any deviation (in matter, gravitational fields, or interference zones) alters the local ATT.

In Summary:

- Light is not an object.
- It does not travel from source to observer.
- The structured vacuum is not empty—it is an active ripple field.
- Perception is not a result of receiving something—it is a result of resonating with something.

Thus, the myth of light as a flow collapses. What we observe is not transit, but alignment.

And this shift in understanding marks the beginning of a new optics, a new energy theory, and a new perception of reality.

2. Wavefronts: The Grand Illusion

In classical wave optics, the concept of a wavefront has been central to explaining how light propagates. Popularized by *Christiaan Huygens* in the 17th century, Huygens' Principle proposed that every point on a wavefront acts as a secondary source of spherical wavelets, which collectively form the new wavefront at a later instant. This elegant idea became a cornerstone of wave theory, especially in explaining phenomena like diffraction and interference.

But there is a problem—a conceptual sleight of hand that went largely unnoticed.

A wavefront, in traditional physics, is not a physical entity. It is defined as a locus of points that share the same phase—a mathematically convenient surface representing a "snapshot" of the wave's state at a given time. But no one ever *sees* a wavefront. It is an imaginary construct, imposed upon space to help us predict behavior based on a presumed flow of energy.

And therein lies the illusion.

Huygens' Principle and the Invention of Loci

To truly appreciate the depth of this conceptual flaw, we must return to how Huygens' principle was framed. The wavefront was never intended to be a substance—it was a geometrical tool, meant to abstractly guide the expected direction of wave propagation. The wave itself was presumed to be spreading in a medium, like ripples on a pond. The wavefront was the leading edge—a visual boundary, like a ripple's crest moving forward.

But light is not a ripple in water. And after the demise of the aether theory, physics conveniently dropped the medium—but retained the abstraction. The wavefront remained in textbooks, equations, and simulations as a ghost of a medium that no longer existed.

Even in modern optics, we still draw these loci: circular or planar surfaces representing equal-time phase states. They help us calculate lens behavior, design instruments, and model interference patterns. But no matter how precise the math, we've forgotten the most basic truth:

A wavefront is not a thing. It is a map of assumptions.

UED Insight: Wavefronts Are Real—But Not How We Thought

Universal Energy Dynamics (UED) turns this picture inside out. Where classical optics treats wavefronts as imaginary guides of an assumed wave motion, UED reveals that what we call a wavefront is actually a ripple structure embedded in the vacuum field.

In UED, there are no waves propagating through a medium.

There are ripples—field-level tensions and deformations—that encode energy and identity.

These ripples are not fictitious or symbolic.

They are consequential distortions in the equilibrium-seeking behavior of the structured vacuum.

And here's the breakthrough:

The so-called "wavefront" is the zone of active ripple alignment, not a passive phase boundary.

It's not a map of where a wave *has reached*; it's the frontline of ripple consequence, where field interaction has reached the threshold for perceptibility or effect.

This completely redefines how we view light behavior.

- In reflection, the ripple boundary resists realignment and redirects tension.
- In refraction, the ripple is forced to reconfigure due to a new ripple density or structural gradient in the medium.
- In diffraction, overlapping ripple fields generate interference zones, where the structured vacuum attempts to resolve multiple simultaneous tension states.

None of this requires phase, frequency, or amplitude in the classical sense. It only requires an understanding of ripple identity, ripple pressure, and alignment zones.

Ripple = Consequence Map, Not Phase Boundary

This insight collapses the old distinction between wave and particle, between motion and interaction. Instead of drawing wavefronts as imaginary surfaces of constant phase, UED invites us to see them as:

- Ripple resonance layers
- Equilibrium-seeking shells
- Energy-consequence zones

The ripple does not move in the way a wave does. It propagates as a pattern of tension seeking alignment. And where that alignment reaches a threshold—what we later define as ATT (Alignment Threshold Time)—a consequence occurs: perception, effect, interaction.

What we thought was a boundary of phase becomes a live ripple map of energetic consequence. The "wavefront" is no longer an artifact of motion—it is a living footprint of interaction potential.

This changes how we interpret every optical phenomenon:

- The angle of incidence is not about path—it is about ripple compression direction.
- Interference patterns are not wave overlaps—they are zones of consequence competition.
- Lensing is not light bending—but ripple contouring by field curvature.

A Final Clarification

In classical models, we've confused the *map* for the *territory*. Wavefronts helped us build mathematical models, but we began to believe in them as realities. In UED, we peel back the metaphor and expose the underlying structure.

So the next time we draw a wavefront in an optical diagram, let us not think of it as a crest of a traveling wave. Let us see it as what it truly is:

A ripple resonance threshold—a snapshot of alignment in action, the very skin of a consequence about to emerge.

This is the real light. Not what flows—but what forms. Not what arrives—but what awakens.

Part II: The Ripple Reality

3. Light as a Ripple Event

In modern physics, we often bounce between two competing models of light: In one breath, it is a wave, spreading across space in crests and troughs; In the next, it becomes a photon, a discrete packet of energy, particle-like and quantized. This duality has become a cornerstone of quantum mechanics—but also one of its greatest paradoxes.

Universal Energy Dynamics (UED) cuts through this paradox—not by choosing a side, but by revealing that both models are incomplete representations of a deeper phenomenon.

Light is not a particle. It is not a wave. It is a ripple event—a structured field activation, not a transmitted object.

Let us reframe light from the ground up—not as a journey, but as an emergence.

No Photon, No Wave: Only Ripple Activation

Photons, as defined in quantum electrodynamics, are quantized excitations of the electromagnetic field—massless carriers of energy and momentum. But these "particles of light" are never directly observed. What we detect in experiments are interactions: a photoelectron released, a chemical reaction triggered, a sensor activated.

This makes the photon not a physical entity traveling through space, but a label for a specific outcome. It is a name we give to a consequence, not a description of a cause.

Similarly, the wave model relies on the assumption that light has a frequency, wavelength, and amplitude traveling through space. But again, these are *inferred properties* based on repeated observations—not direct features of a thing in motion.

UED bypasses this confusion by introducing a more fundamental model: Light is not a thing in motion—it is a field event. Specifically, a ripple activation within the structured vacuum.

What we interpret as a photon or a wave is simply the field's localized response to a ripple reaching alignment threshold—what we now define as ATT (Alignment Threshold Time).

Structured Vacuum and Ripple Dynamics

The vacuum is not empty. According to UED, space is filled with a structured vacuum field—an active, organized lattice of potential ripple interactions. This field is not static or chaotic; it is governed by equilibrium-seeking behavior. Every energetic event introduces tensions in this field. These tensions are not waves in the classical sense—they are ripples, consequences radiating outward in search of balance.

Ripples in UED carry no substance. They do not transport photons. Instead, they carry:

- Pattern identity
- Energy potential
- Alignment demand

These ripples move through the vacuum not by displacing anything, but by restructuring the equilibrium of the field itself. In other words, they change the conditions under which a consequence may arise.

The Role of the Source: Triggering Ripple Distortion

When a light source emits energy—whether a heated filament, an electron transition, or nuclear decay—it distorts the local structured vacuum. This disturbance is not an emission of a "thing," but a ripple signature: a pattern of imbalance radiating outward.

This signature doesn't push particles across space. Instead, it propagates alignment pressure—a kind of informational tension—through the ripple network of the vacuum.

The ripple field carries no photons. What it carries is an invitation to align.

And here is the key insight: This ripple continues through space until a structure—an eye, a sensor, a leaf—enters a state capable of matching its ripple pattern. That's when the consequence occurs.

The Role of the Observer: Aligning to Ripple Signature

The observer is not a passive receiver. In UED, the observer is an active participant in ripple alignment.

Whether it's the retina of a human eye or the silicon of a solar panel, the detecting structure offers a local ripple configuration. When a ripple signature introduced by the source matches or resonates with this local configuration, an activation occurs—an electron is displaced, a neuron fires, a perception emerges.

No particle was received. No wave traveled across space. The observer simply reached alignment threshold—and consequence followed.

This is illumination: Not a receipt of energy—but a field consequence of ripple harmony.

Implications of the Ripple Event Model

This model explains numerous observations without invoking paradoxical dualities:

- Why light appears particle-like in photoelectric experiments: → Because ripple alignment is discrete—either a system resonates, or it doesn't.
- Why light appears wave-like in interference: → Because ripple fields interact, producing complex alignment zones—not overlapping amplitudes, but competing ripple pressures.
- Why "detection" is always interaction-based: → Because nothing travels to the observer; the observer becomes the consequence.

This also dissolves the question: *"Where is the photon before detection?"* It never existed. There was only the ripple. Detection is the photon, not its endpoint.

Ripple Activation Is Not Motion, But Resolution

The most transformative implication of the ripple model is this: Light is not about moving from place to place—it is about resolving tension in the vacuum field.

Once a ripple is triggered, it becomes a pattern of possibility in the structured vacuum. It does not carry energy as substance, but as pressure for alignment. And wherever this alignment is achieved—perception, energy transfer, or interaction results.

This shifts optics from kinematics to consequence dynamics. It shifts light from being an object to being a ripple event in time and structure.

Closing Reflection for This Section

To call light a ripple is not to diminish it. It is to finally understand it—not as something that "shines," but something that summons. Not something that "travels," but something that realigns the universe in favor of recognition.

We don't detect light. We become the condition under which light is possible.

And that changes everything.

4. The Nature of Illumination

We are so accustomed to associating illumination with light arriving, entering, and lighting up our world that we rarely question what is really happening when something becomes "visible." Light, in everyday understanding, is assumed to be a stream of particles or waves that "hit" our eyes, sensors, or surfaces—activating a response.

But under Universal Energy Dynamics (UED), this common view crumbles.

Illumination is not a process of reception. It is not about catching light as if it were a thrown object. It is a field-level consequence of ripple alignment—a structural interaction between energy identity and perceptual readiness.

This section redefines illumination not as a passive result of incoming radiation, but as an active outcome of alignment thresholds being met within the structured vacuum between a source and an observer.

Perception Is Not Reception

To receive something is to be passive. To perceive something is to be active.

Traditional optics assumes that light arrives—carrying packets of energy or vibrating wavefronts—and the observer simply receives it. This model treats the eye, camera, or sensor as a kind of bucket, waiting to collect these "light objects."

But perception isn't passive bucket-filling. It's active field engagement. In UED, the observer is not a receiver of light—they are a participant in ripple resolution.

Your eye, for instance, doesn't just wait for light to come in. Its photoreceptors exist in a ripple-ready configuration—with molecular structures poised to align with specific energy patterns in the structured vacuum.

When alignment occurs, perception is triggered. Not because a photon arrived, but because the ripple field reached a moment of matching—and this matching itself causes the perception.

So instead of thinking "light reached my eye," UED invites us to say:

"My local field aligned with the ripple signature of the source, and perception was born."

This is a profound shift. It moves perception from being reactive to being interactive.

Illumination as Field-Level Consequence

Illumination is not an object shining into space. It is a ripple effect becoming visible because the structured vacuum permitted it—because the field around the observer resolved into that specific energy pattern.

The ripple introduced by a source does not go searching for an eye. It simply modifies the field condition in all directions, creating zones where alignment is possible.

Only when a structure with compatible ripple configuration enters that zone—like your retina or a camera sensor—does the ripple's potential resolve into a consequence. That consequence is what we call illumination. So illumination is not caused by the *arrival* of light. It is the activation of field harmony.

In this view:

- A glowing bulb is not throwing photons.
- It is creating a persistent ripple disturbance in the surrounding vacuum.
- That ripple becomes illumination only where structures capable of resolving it are present.

You do not "see" because light hits you. You see because your field aligned with a ripple pattern that was already pressuring the vacuum to express something.

Ripple Alignment and Detection: The Real Trigger

This brings us to the crux of the phenomenon: What we call detection—be it the firing of a rod cell in your eye, or the triggering of a pixel in a CCD sensor—is not caused by receiving a photon or energy quantum. It is caused by successful ripple alignment.

Here's how it works in UED terms:

1. The source event (e.g., atomic transition) introduces a ripple distortion into the structured vacuum.
2. This ripple carries no mass or momentum—it carries identity, intent, and pressure for alignment.
3. This ripple modifies the surrounding field into a zone of alignment tension—a ripple resonance region.
4. When a detecting structure enters this region and offers a compatible ripple profile (chemical, geometric, or energetic), alignment threshold is met.
5. Detection occurs—not because something "hit" the detector,

but because ripple harmonization reached consequence level.

This detection—this spark of illumination—is therefore the real event. The so-called "photon" never existed; it was only the label we applied to this ripple alignment outcome.

Examples from Everyday Illumination Reinterpreted

Let's consider a few common scenarios to illustrate this ripple-based model of illumination:

A candle in a dark room: In traditional thinking, photons radiate outward. In UED, the flame modifies the local field. Everything "visible" in the room is not lit because photons hit it, but because your eyes' field reached ripple alignment with reflected or emitted ripple structures.

Laser beam in fog: What you see is not a laser "passing through" fog. You see zones where ripple alignment with water droplets momentarily exceeds threshold, triggering visual consequence.

Sunlight entering your eye: You're not catching sunlight. Your retina is aligning with ripple patterns in the vacuum field already activated by the sun's emissions. Your perception is the local resolution of that global field tension.

Consequences of This Reframing

This reframing leads to a series of revolutionary implications:

- There is no need for a "medium" for light to travel. The vacuum itself, as a structured ripple field, is sufficient for ripple propagation and alignment.
- Speed of light becomes irrelevant. What matters is ATT—Alignment Threshold Time, which is a measure of

how fast alignment is achieved, not how fast something moves.

- Detection becomes a field event, not a collision event. This redefines photodetection, energy absorption, and even biological vision at the root.
- No duality exists. Light is neither wave nor particle. It is ripple pattern seeking consequence through structured vacuum interaction.

Closing Thought: Seeing Is Resonating

Illumination is not a thing. It is a moment— A moment when the universe, through your local structured field, says: "Yes. You are ready to see." Light doesn't reach you. You reach the ripple. And together, you become the consequence. That is illumination.

5. Stealth by Misalignment

A UED-Based Approach to Invisibility

Invisibility is not the absence of existence—it is the absence of alignment. In traditional physics, stealth technologies aim to reduce detection by minimizing reflection, absorption, or thermal output. But under UED, stealth is redefined as a ripple interaction problem, not an energy suppression problem.

When radar or infrared sensors attempt to detect an object, they rely on ripple resonance—the alignment of emitted or reflected ripple patterns with the detector's field configuration. If no alignment occurs, no ATT (Alignment Threshold Time) is reached, and the object remains undetected—not because it is invisible in the conventional sense, but because no consequence is triggered in the sensing system.

How Ripple-Based Stealth Works:

Field Disruption Avoidance: The object is designed to avoid introducing sharp ripple distortions into the surrounding vacuum.

Ripple Signature Mismatch: The object's material or geometry distorts or redirects ripples such that they do not align with standard sensor ripple expectations.

Adaptive Field Shaping: The aircraft or drone actively manipulates its boundary field to ensure ripple pressure diffuses without matching any nearby detection system's field configuration.

Applications:

Radar Evasion: Aircraft configured to prevent ripple echo in radar bands become functionally undetectable.

Thermal Camouflage: Ripple damping in infrared frequencies prevents sensor lock.

Visual Invisibility: Avoidance of ripple reflection in the visible band leads to practical cloaking.

Strategic Shift:

Instead of trying to absorb or deflect energy, UED-based stealth ensures that the object never enters the ripple resolution range of any observer.

Insight:

"To be invisible is not to be absent. It is to escape alignment."

UED offers not just a new optics—it offers a new defense philosophy. Objects designed with ripple misalignment in mind can bypass detection across all domains—optical, electromagnetic, and even acoustic. This isn't low visibility. This is non-consequence engineering.

Part III: Introducing ATT – Alignment Threshold Time

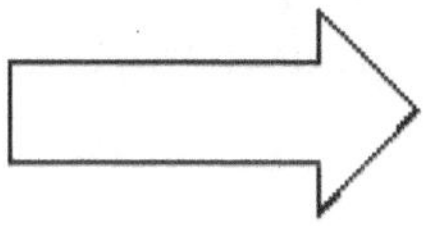

6. ATT Explained – Alignment Threshold Time

If there is one cornerstone concept that emerges from the Universal Energy Dynamics (UED) perspective on light, it is this: Alignment—not movement—is the source of perception. And if alignment is the trigger, then the time it takes to achieve that alignment becomes more fundamental than any notion of "speed."

That time is what we define as ATT—Alignment Threshold Time.

In this section, we formally introduce ATT as a core metric in UED-based optics, contrast it with the traditional idea of light-speed, and explore the variables that influence its value.

ATT Defined: The Time Needed for Ripple Alignment

ATT is the minimum temporal interval required for a ripple disturbance—introduced by a source—to reach consequence-level alignment with an observer or detector's local structured vacuum configuration.

In simple terms:

ATT is the time it takes for "something to become visible" through ripple resolution.

It is not the time for an object to travel. It is the time for a field condition to harmonize enough that it triggers a detectable outcome—whether visual, thermal, chemical, or energetic.

This changes everything.

- ATT is not about distance.
- ATT is not about transmission.
- ATT is about field-level readiness.

In fact, light-speed (c) in traditional physics can now be seen as a special-case ATT in ideal vacuum—the fastest possible field alignment in a non-turbulent, equilibrium-ready medium.

But in reality, ATT is contextual and variable.

Variables that Influence ATT

ATT is not a universal constant. It changes depending on the properties of the field between the source and the observer. Key factors include:

1. Ripple Density

- The structured vacuum is never truly empty.
- In high ripple-density environments (e.g., near energetic bodies, inside matter), existing tensions compete with new ripple patterns.
- Result: Longer time to achieve clean alignment → increased ATT.

2. Turbulence in Field Structure

- Any medium (air, water, plasma) adds structural irregularities to the ripple field.
- These distortions deflect, delay, or decohere the incoming ripple pattern.
- ATT increases in proportion to field irregularity.

3. Resonance Zone Availability

- Alignment requires not just pattern similarity but resonant compatibility.
- If the observer's local field is mismatched—by geometry, energy state, or configuration—alignment takes longer or may not occur at all.
- Think of it like tuning a radio: if you're not near the right frequency, the signal doesn't resolve.

4. Material Ripple Refractive Index

- Traditional optics speaks of refractive index as a ratio of light speed.
- UED redefines it as ripple resonance delay index—how much longer a ripple takes to align due to the material's internal structure.
- Every material has a different ATT profile depending on its ripple propagation compatibility.

ATT vs. Speed of Light: A New Metric, A New Paradigm

The speed of light (c)—299,792,458 m/s—is often treated as the most sacred constant in physics. It's written into equations, used to define units, and serves as the ultimate speed limit in Einstein's universe.

But in UED, "c" is not a velocity. It is an illusion of minimal ATT in vacuum.

Let's compare the two:

Feature	Classical Light Speed (c)	ATT (Alignment Threshold Time)
Based on	Distance/time	Field alignment interval
Assumes	Light as a particle/wave	Ripple as consequence trigger
Constant?	Yes (in vacuum)	No – context and structure dependent
Interprets medium as	Obstacle to velocity	Modifier of ripple resolution
Usefulness	Good for engineering models	Better for understanding perception
Foundation	Kinematics	Consequence dynamics

ATT doesn't break physics—it clarifies it. It removes the unnecessary mental gymnastics of imagining light traveling as a substance, and replaces it with a more elegant and accurate model:

What we detect is not what arrived. It's what aligned.

Illustrative Analogy: Radio Tuning vs. Postal Delivery

- The traditional view treats light like mail delivery: it starts at the post office, travels down roads, and arrives at your door.
- ATT reframes it as radio tuning: the moment your receiver is aligned, you hear the station—not because the wave "arrived," but because your system matched the field condition that already existed.

ATT, therefore, isn't limited by distance, but by resolution capacity.

Implications Across Science

- Time Dilation becomes ripple friction misinterpreted as time stretch.
- Optical Fiber Delay becomes ATT variation—not energy

slowness.

- Black Hole Event Horizon = infinite ATT zone—ripple alignment is no longer achievable for outside observers.
- Quantum Tunneling = ripple alignment bypassing structural barriers—not particles "passing through."

Closing Insight: ATT as the Rhythm of Reality

We measure seconds, meters, frequencies, and energies. But beneath all of it is a single question: When does the field say "Yes"? ATT is the rhythm of that yes. It's not the time something takes to reach us. It's the time we take to match what was already there. And that, perhaps, is a more honest definition of perception.

7. Ripple Behavior vs. Speed

In classical physics, speed is one of the most fundamental quantities—defined simply as the rate of change of position with time. We speak of light "traveling" at 299,792,458 meters per second, as if it were a car on a cosmic highway, moving smoothly from one point to another.

But under the Universal Energy Dynamics (UED) framework, this entire picture unravels. Because if there is no "thing" traveling, then what exactly is moving? What is it that possesses speed?

The answer: nothing in the way we thought.

In UED, light is not a thing moving—it is a ripple event occurring. And so, speed becomes a projection—a mathematical artifact of how we interpret timing between events—not a causal, physical property.

This section explores how ripple propagation operates fundamentally differently from classical movement, and why ATT—not speed—is the true measure of perception.

Speed Is a Projection, Not a Cause

Let's begin by addressing a foundational truth: Speed doesn't cause anything.

We often speak as if things happen *because* of their speed—bullets penetrate due to speed, meteors burn up due to speed, light reaches us due to speed.

But speed is not a force. It is not a mechanism. It is a derivative—a ratio. It tells us nothing about the underlying cause of a phenomenon. Only how fast it appears to unfold relative to an observer's clock and ruler.

When we measure the "speed of light," we are not observing a thing moving. We are measuring a time interval between a source's ripple activation and the observer's response—and assuming that something "traveled" between them.

In truth, we are measuring how quickly alignment is achieved—and interpreting that as a "velocity."

This is what UED corrects.

Ripple Propagation Is Tension-Release Behavior

In the structured vacuum of UED, ripples do not move like objects. They propagate more like tension-seeking harmony spreading across a field of potential. Picture the structured vacuum like a cosmic fabric—not empty space, but an intelligent medium full of equilibrium patterns.

When energy disturbs this field, a ripple is born—not a wave of motion, but a disturbance of consequence. This ripple does not carry matter or mass. It carries:

- Pattern Identity
- Field Intent
- Resolution Demand

The ripple's goal is not to *reach a destination*, but to resolve tension across distance. Its behavior resembles a chain reaction, a pressure redistribution, a field-level call for balance.

This tension-release process does not depend on velocity. It depends on:

- How structured the field is
- How much interference exists
- How compatible the ripple signature is with potential alignments

In a perfect vacuum (uniform ripple field, no interference), this tension resolves very quickly—and we observe what we call "light-speed." But in other contexts (dense matter, gravitational distortion, misaligned detectors), ripple resolution slows down—not because light is moving slower, but because alignment takes longer.

ATT Determines Perception—Not "How Fast" Light Travels

This is where the Alignment Threshold Time (ATT) reclaims center stage.

Perception—whether biological or mechanical—does not occur when "light arrives." It occurs when a ripple pattern aligns with the local field configuration enough to cross the detection threshold.

That threshold can be reached:

- Almost instantly (ideal conditions, clear path)
- Slowly (through dense or turbulent fields)
- Never (if alignment is impossible or forbidden)

So, what we *think* of as speed—light slowing in water, bending through glass, or being delayed near black holes—is actually:

ATT varying in response to local ripple complexity and structure.

This resolves many classical anomalies:

- **Refraction**: Longer ATT in higher-density ripple zones, not slower velocity.
- **Fiber optic delay**: ATT distortion across multiple alignment layers.
- **Black hole event horizon**: ATT tends toward infinity; no consequence ever resolves beyond threshold for external observer.

Ripple Is Not a Carrier. It's a Coordinator.

Ripples are not vehicles. They are field-level communicators—coordinating changes across space, not ferrying particles.

So we must abandon the image of light as a car zooming across the highway of space. Instead, visualize a structured vacuum trembling with tension, and ripples as signals within this field, coordinating energy realignment.

In that picture:

- The field is primary.
- The ripple is the intent.
- ATT is the resolution moment.
- And speed? Just an illusion born of distant clocks and desperate equations.

Closing Reflection: From Speed to Consequence

Speed helped us describe. ATT helps us understand.

The world is not made of things in motion. It is made of ripples in resolution. You never "see light." You simply exist at the moment when the field says "now."

And that's not about motion. It's about consequence.

Part IV: Revisiting Classic Experiments

8. Michelson-Morley Reinterpreted

A Ripple-Based Explanation

The Michelson-Morley experiment, conducted in 1887, is one of the most famous "null results" in the history of science. Designed to detect the Earth's motion through the hypothesized luminiferous aether (the supposed medium through which light waves travel), the experiment found... nothing. No difference in light speed based on direction. No aether wind. No delay where one was expected.

This result shook physics. It led to the rejection of the aether theory and became a precursor to Einstein's Special Relativity. Yet, even in abandoning the aether, physics retained a more deeply rooted illusion:

That light still flows—just now, through "empty space."

In this section, we reinterpret the Michelson-Morley experiment not through the lens of flow, but through the UED framework of ripple alignment. What the experiment failed to detect wasn't a failed theory—it was a misunderstood phenomenon.

No Aether Needed—Because No Flow Is Assumed

The original hypothesis of the Michelson-Morley setup assumed that:

1. Light is a wave that travels through a medium (the aether).
2. The Earth, moving through this medium, would cause a relative "aether wind."
3. This wind would create measurable differences in light's travel time along perpendicular arms of an interferometer.

When no difference was observed, physicists were forced to abandon the idea of a static aether. But they never abandoned the idea of light as a traveling entity.

Einstein's solution was to fix the speed of light as a universal constant, introducing time dilation and length contraction to preserve the mathematics of motion—but never questioning whether light actually "moves" through space at all.

UED breaks this final illusion. It doesn't just discard the aether—it discards the very idea of flow-based light.

In UED:

- There is no medium required, because light is not a transmission.
- There is no flow, because light is not a traveler.
- There is only ripple propagation and alignment.

What Really Happened: Delay in Ripple Alignment, Not Transit Time

So what did Michelson and Morley actually measure?

Their interferometer sent beams of light along two perpendicular arms, bounced them off mirrors, and recombined them. Any difference in travel time was expected to cause a shift in interference patterns due to the Earth's motion through aether.

They found none.

But under UED, this outcome is not mysterious at all. Here's what was really happening:

- The light source introduced a ripple pattern into the

structured vacuum.

- This ripple propagated not as a wave in a medium, but as a tension-seeking distortion in the vacuum field.
- The mirrors and detectors aligned with this ripple pattern when their local field configurations reached a threshold of ripple compatibility.
- The time for detection was not dependent on the path length, but on how quickly ripple alignment was achieved at each point in the setup.

Since the entire apparatus was moving as a ripple-coherent system through the structured vacuum, both arms experienced equivalent alignment conditions, despite being oriented differently. The field was already structured to support symmetric ripple behavior, meaning no differential ATT (Alignment Threshold Time) emerged.

This is why the outcome was null—not because the aether doesn't exist—but because the measurement assumed something that wasn't real: transit time.

Why the Null Result Is Expected in Ripple Dynamics

UED shows that a "null result" in the Michelson-Morley context is not an anomaly—it's inevitable.

Ripple propagation does not follow a vector like wind or sound in air.

It is a field-wide restructuring event, governed by:

- Ripple pattern fidelity
- Vacuum field uniformity
- System symmetry

Since the experimental apparatus is coherent within the same ripple context, its components resonate with the same field structure, regardless of orientation or motion through space. As a result:

- No phase difference appears
- No alignment lag occurs
- No directional ripple delay manifests

The ripple is not moving through something—it is restructuring everything it touches simultaneously, based on pattern compatibility. This includes mirrors, arms, detectors, and the source itself—all ripple-linked in one coherent field experience.

The Ripple Advantage: No Need for Correction Factors

Relativity had to introduce Lorentz transformations, time dilation, and length contraction to explain why light speed remains constant across frames.

UED needs none of these.

- No velocity of light to preserve
- No space-time bending to protect equations
- Just field-based alignment and ATT variation, rooted in physical consequence

The structured vacuum naturally equalizes ripple alignment conditions, removing the need for compensatory geometry or metaphysical postulates.

Closing Thought: What They Didn't Measure Was the Truth

Michelson and Morley believed they were measuring how fast light traveled along two arms of an interferometer. But what they were unknowingly observing was the constancy of ripple alignment in a structured field system.

Their null result didn't mean "nothing happened." It meant that their foundational assumption—that light flows—was flawed. And now, over a century later, we can finally give that null result its proper interpretation: They failed to detect transit time... because there was none to detect.

Just a ripple. And a field. Waiting to align.

9. Optical Phenomena, Now Clear

Refraction, Reflection, and Diffraction in Ripple Dynamics

Classical optics offers a robust set of mathematical rules for how light behaves—Snell's Law for refraction, the law of reflection, and wave interference principles for diffraction. These laws work well in practice, but they are built on a model that assumes light is a wave or a particle traveling through or bouncing off surfaces.

However, in the Universal Energy Dynamics (UED) framework, none of these interactions are about travel, collision, or wave propagation in the classical sense. They are all about ripple alignment, boundary behavior, and field-level consequence.

This section recasts these foundational phenomena—refraction, reflection, and diffraction—not as outcomes of trajectory and energy transfer, but as field expressions of ripple tension and alignment thresholds. Once the illusion of flow is removed, these optical behaviors become not only clearer, but more fundamentally coherent.

Refraction = Ripple Realignment

In classical physics, refraction is explained as the bending of light when it passes from one medium to another (e.g., air to water). The cause? A change in velocity, based on the material's refractive index. This model relies on light slowing down in denser media—an odd idea, given that photons are massless and should, in principle, always move at "c."

In UED, refraction is not caused by a change in speed. It is caused by a forced realignment of ripple propagation within a new structured field configuration.

Let's break that down:

- Every material has a unique ripple structure—a set of ripple propagation pathways based on its atomic geometry, density, and vacuum tension profile.
- When a ripple enters a new medium, it encounters a different ripple lattice, forcing it to reconfigure its alignment path.
- This reconfiguration appears as "bending," but what's actually happening is a change in the ripple tension's direction of least resistance.

Instead of asking, "How fast does light go here?" we ask:

"How does the ripple pattern realign to maintain equilibrium in this new field?"

This realignment explains:

- Why the angle of refraction changes predictably (Snell's Law becomes a ripple equilibrium law)
- Why no energy is lost (just re-channeled)
- Why the ripple doesn't "slow down," but adjusts its ripple pathway geometry

In essence, refraction is ripple path realignment in search of structural resonance.

Reflection = Ripple Boundary Resistance

Classically, reflection is described as light bouncing off a surface, where the angle of incidence equals the angle of reflection. It's a geometric explanation assuming the light ray behaves like a billiard ball.

UED reframes reflection as ripple rejection—an event where the ripple's structure fails to penetrate or realign with the boundary field of a material and is thus deflected by ripple tension resistance.

In detail:

- The surface of a reflective material presents a boundary ripple configuration that is incompatible with the incoming ripple.
- This mismatch creates ripple pressure buildup, which cannot resolve within the surface.
- Instead, the ripple redirects—not as a bounce, but as a field tension redirection event.

This explains:

- Why reflection preserves the angle of incidence (ripple vectors re-emerge at the minimal distortion path)
- Why some surfaces are more reflective (they have tighter, less permeable ripple boundaries)
- Why total internal reflection occurs—not due to angle alone, but due to ripple confinement and containment

So, reflection is not a mirror throwing light back—it is the field saying:

"This ripple pattern does not belong here. Redirecting."

Diffraction = Ripple Interference and Field Fatigue

In traditional wave optics, diffraction is described as light bending around obstacles or spreading after passing through a narrow slit. It's treated as a side effect of wavefront curvature and wave interference.

But in UED, diffraction is not about wavefront spreading. It is a ripple interaction zone, where competing field structures lead to interference tension and vacuum fatigue.

Here's what really happens:

- A ripple passing through a narrow slit is forced into a constrained ripple geometry.
- The structured vacuum near the edge of the slit experiences high ripple pressure, which leads to field strain.
- This strain creates secondary ripple disturbances—not spherical wavelets (as in Huygens), but local field fatigue zones that interfere with the main ripple path.
- These interactions generate ripple pattern divergence, producing what we observe as spreading, bands, or dark/light fringes.

Importantly, this is not "light bending." It is ripple reexpression due to spatial constraint and energetic redistribution.

This also explains:

- Why sharper edges create more pronounced diffraction: they generate stronger ripple shear
- Why different wavelengths diffract differently: longer ripple patterns are more vulnerable to edge-induced field fatigue
- Why diffraction patterns self-heal over distance: the field gradually rebalances ripple tension

Thus, diffraction is not a wave effect, but a field consequence of ripple confinement and misalignment near structural boundaries.

Unified Field Insight: These Phenomena Are Not Separate

From the ripple dynamics perspective:

- Refraction, reflection, and diffraction are not different phenomena.
- They are different expressions of how ripple fields respond to structure.

Classical Term	UED Reinterpretation
Refraction	Ripple path realignment due to field shift
Reflection	Ripple resistance at field boundary
Diffraction	Ripple interference via edge fatigue

They all share a single origin:

Ripple behavior under structural tension.

No flow. No speed change. No duality.

Just structured consequence, beautifully playing out across reality.

Closing Reflection: The Language of Light is Ripple

These everyday optical phenomena are not mysterious, nor are they magical abstractions requiring waves and particles to coexist paradoxically. Under UED, they are the natural outcomes of ripple alignment within structured vacuum fields.

They are not illusions—but neither are they what we once thought.

The light that bends, reflects, or splits is not moving. It is resolving tension.

It is translating structure into consequence.

And now, for the first time, we can see it not as trickery—but as truth.

Part V: Beyond Optics – Impacts Across Physics

10. Collapse of Wave & Ray Optics

A Ripple-Based Redefinition

Wave optics and ray optics have been the two dominant models in explaining how light behaves in space, matter, and instruments.

- Wave optics interprets light as a continuous electromagnetic wave, enabling explanations of interference, diffraction, and polarization.
- Ray optics simplifies light as straight-line trajectories, useful for modeling reflection, refraction, and image formation in lenses and mirrors.

Both models work well—until they don't. Each has limitations, paradoxes, and inconsistencies. Together, they often require switching mental models mid-analysis, invoking one metaphor or another depending on context.

Universal Energy Dynamics (UED) dissolves this divide. In UED, there are no waves, no rays—only ripple dynamics within a structured vacuum.

What emerges is a unified, consistent model where both wave and ray behaviors are reinterpreted as manifestations of ripple-based field interactions. This is not a hybrid theory—it is a collapse of the old dichotomy into a more foundational reality.

Wave Optics Becomes Ripple Interference Theory

In wave optics, interference patterns result from overlapping waves that reinforce or cancel each other based on phase alignment. However, this relies on abstract constructs: infinite wavefronts, superposition assumptions, and probabilistic interpretations in the quantum domain.

Under UED, we don't have overlapping waves. We have ripple fields, where multiple ripple patterns create zones of tension and harmony within the structured vacuum.

Here's what changes:

- Interference is no longer "wave + wave = amplitude." It is ripple pattern interaction, where tensions in the vacuum reinforce or cancel based on compatibility.
- Constructive interference occurs where multiple ripple pressures align toward the same consequence.
- Destructive interference occurs where ripple tensions neutralize, relieving the field without triggering consequence.

This eliminates the need for imaginary wavefronts, superposition metaphors, or probability clouds. What we see as interference patterns are the visual echoes of ripple tension resolution zones—real field phenomena, not wave illusions.

So, wave optics collapses into:

Ripple interference theory – a physical interaction of field-level consequence, not abstract amplitudes.

Ray Optics Becomes Ripple Path Optimization

Ray optics assumes light travels in straight lines until something bends it—via reflection or refraction. Rays are imaginary paths we draw to model where light goes and how images form. But rays don't exist. They are projections, not properties of light itself.

In UED, light doesn't move along a path. It doesn't travel. What we perceive as a "ray" is the optimal path of ripple alignment—where the field permits the fastest or most efficient ripple resolution.

- A ray is not a thing. It is the path of minimum ATT (Alignment Threshold Time) through a structured field.
- Mirrors don't "bounce" light; they redirect ripple tension along the fastest return path.
- Refraction bends rays not because of velocity changes, but because the ripple seeks the path of least field resistance.

So, ray optics becomes ripple path optimization:

- Rays = field alignment contours
- Focus points = convergence zones of ripple identity
- Caustics = complex ripple tension maps in curved fields

This shifts optics from geometric modeling to consequence modeling—far more physically grounded.

All Lens Behavior = Field Shaping

In both wave and ray optics, lenses are treated as devices that either bend rays or change wave phase. But neither explanation describes what a lens actually is, or why it works at the most fundamental level.

In UED, a lens is not bending anything. It is reshaping the structured vacuum field to guide ripple alignment.

Let's break this down:

- The curvature of the lens alters the ripple propagation environment by modifying the density and geometry of the ripple-compatible field.

- As ripples enter this region, their pathways are constrained and redirected toward a focal zone—not because they are bent, but because the field demands reconfiguration for alignment.
- Focal points are not the intersection of rays—they are the lowest ATT zones for that ripple signature within the field.

This explains:

- Why lenses of different shapes produce different focal behaviors: they are modulating ripple field geometry, not directing rays.
- Why aberrations occur: imperfect ripple shaping leads to inconsistent ATT across the field.
- Why lenses work even in non-visible wavelengths: because the ripple model is field-based, not wavelength-bound.

So, lens behavior is not geometry-based ray tracing, but structured vacuum sculpting for ripple convergence.

Unified Model: Ripple Dynamics Over Dualism

What used to require two separate models—waves and rays—now resolves into a single, elegant framework:

- There are no rays.
- There are no waves.
- There are only ripples in structured vacuum seeking field equilibrium.

Classical Concept	UED Interpretation
Wavefront	Ripple resonance boundary
Ray	Optimal ripple alignment path
Lens	Field-shaping geometry for ATT minimization
Focus	Consequence convergence point
Interference	Field consequence zones from ripple conflict
Reflection	Ripple rejection and redirection
Refraction	Ripple path realignment

This collapse of wave and ray optics into ripple dynamics is not a simplification—it's a liberation from metaphor.

UED doesn't just explain more. It explains why the previous models were ever needed—because we lacked a model of consequence-based interaction until now.

Closing Insight: The Eye Sees the Ripple, Not the Ray

In the final analysis, neither rays nor waves were ever real. They were metaphors—useful, but limited. Projections, not properties.

The lens in your eye doesn't bend rays. It modulates ripple readiness across your retina. What you see is not a ray's path, but a ripple consequence shaped by your biological field.

And now, with UED, we finally have a model that speaks the language of nature: Field, tension, alignment... and perception.

11. Nuclear Energy Reframed

Ripple Restructuring in Atomic Events

In the conventional view of nuclear physics, energy is said to be released through the conversion of mass—either by splitting heavy atoms (fission) or fusing lighter ones (fusion). Photons, particularly gamma rays, are treated as high-energy packets of electromagnetic radiation "emitted" during these transitions.

But this model, as widely accepted and mathematically successful as it is, is built upon a substance-centric paradigm:

- Energy is "stored" in mass
- Photons "carry" energy away
- Particles and nuclei act as containers and projectiles

Universal Energy Dynamics (UED) offers a fundamentally different lens. It views nuclear processes not as particle events, but as ripple restructuring events within the structured vacuum. The nucleus is not a box of particles—it is a localized field configuration, and any nuclear transition is an act of ripple reconfiguration under extreme structural pressure.

Let's walk through the new model.

Emission Is Ripple Restructuring, Not Photon Escape

In the standard model, when a nucleus undergoes a transition (e.g., radioactive decay or electron capture), it is said to emit a photon—specifically a gamma ray. This is visualized as a high-energy particle of light flying out of the atom.

But in UED, no photon is ever emitted—because no photon is ever real in transit.

What happens instead:

- The nuclear configuration introduces a ripple distortion in the local structured vacuum due to its internal field instability.
- This ripple is not "shot outward" as a particle—it propagates as a consequence-seeking field realignment.
- When external matter (like a detector) aligns with the ripple's signature, an activation event occurs, which we interpret as photon detection.

So, what we've long treated as "photon emission" is better understood as:

A ripple consequence of internal field restructuring reaching alignment with an external field configuration.

This explains why:

- Emission appears discrete (ripple thresholds reached)
- Energy seems "transferred" (field tension resolved into activation)
- Directionality exists (ripple tension paths are geometry-dependent)

No photon flew. A field resolved—and something happened where alignment allowed.

Gamma Rays as High-Frequency Ripple Bursts

Gamma radiation is typically described as electromagnetic waves of very short wavelength and very high frequency, often produced in nuclear reactions. They're labeled as "photons," but unlike visible light, they're associated with extreme energies.

UED simplifies and sharpens this understanding: Gamma rays are not particles nor classical waves. They are concentrated, high-pressure ripple bursts—ripples of the highest structural tension.

In ripple terms:

- High-frequency = rapid field modulation, not literal oscillation
- Short wavelength = tightly packed ripple signature, leading to localized consequence
- High energy = extreme ripple field strain, forcing aggressive realignment across any receptive medium

This makes gamma radiation:

- More penetrating, not because it's "stronger," but because it carries more field-disruptive ripple geometry, which aligns with deeper atomic layers
- More damaging, because it doesn't "hit"—it forces structural field compliance, breaking molecular alignments

This model resolves paradoxes in radiation shielding, material response, and quantum unpredictability. The interaction is not about energy absorption. It's about:

Ripple coherence overload in atomic structures.

Fission and Fusion: Field Discontinuity Events, Not Mass-Energy Games

Traditional nuclear physics explains fission and fusion as mass-to-energy conversion processes, per Einstein's famous equation $E = mc^2$. In this view:

- Fission splits a heavy nucleus into smaller ones, releasing energy due to reduced binding mass
- Fusion combines light nuclei into a heavier one, releasing energy when the resultant nucleus has lower mass than the sum

But this framework assumes mass is a reservoir, and energy is liberated when mass changes. It also forces us to treat mass as substance, rather than a field condition.

UED offers a different truth:

Mass is not energy. Mass is ripple complexity. Energy is ripple resolution.

Fission in UED:

- A heavy nucleus has a complex ripple configuration—internally stable but externally pressured
- When distorted beyond its alignment threshold, the internal field snaps, causing a ripple field discontinuity
- This releases ripple tension in all directions—what we observe as radiation, kinetic energy, and "binding energy" release
- The daughter nuclei are simpler field structures, requiring less ripple containment

So, fission is not about mass loss—it's about field simplification and ripple unbinding.

Fusion in UED:

- Light nuclei (e.g., hydrogen isotopes) each have their own ripple identity
- Under extreme pressure and temperature, these identities are forced into alignment, creating a new compound ripple field
- The merging field releases excess ripple pressure as alignment resolves—what we interpret as energy
- The "mass deficit" is merely a change in ripple containment requirement, not substance annihilation

Thus, fusion is not combining matter. It is ripples merging into a more stable consequence pattern, with tension released along the way.

Field-Level Summary

Classical Concept	UED Interpretation
Photon Emission	Ripple restructuring alignment event
Gamma Radiation	High-frequency ripple burst
Nuclear Fission	Field discontinuity with ripple simplification
Nuclear Fusion	Ripple merging with tension release
Energy Release	Ripple realignment consequence
Mass Deficit	Ripple containment variance

Closing Reflection: The Atom Is Not a Container—It Is a Ripple Nexus

The atomic nucleus is not a dense pocket of energy. It is a ripple node—a structured configuration in the vacuum where ripple tension is intricately bound. Every nuclear event is not an explosion of matter—it is a collapse or unification of ripple identity.

What emerges as heat, light, or radiation is not energy leaving a system. It is alignment echo—the field rebalancing itself after being momentarily broken.

Nuclear energy, then, is not power extracted from matter. It is the structured vacuum resolving its own imbalance.

Part VI: Practical and Experimental Horizon

12. Measuring ATT – A New Science

If ATT (Alignment Threshold Time) is to replace the concept of light-speed as a more foundational physical measure, it must not remain theoretical. It must be measurable, testable, and practically applicable. Just as Galileo's inclined planes made acceleration observable, and Michelson's mirrors attempted to pin down the speed of light, we now stand at the threshold of building the first generation of ATT-based instruments—designed not to detect velocity, but to capture ripple alignment as a time-bound consequence event.

This section outlines the framework for such experimentation, explains how ATT varies across different media, and addresses fascinating ripple-based anomalies such as instantaneous perception and blackout zones—long misunderstood or ignored by mainstream optics.

Proposed Setup: Detecting Alignment Thresholds

To measure ATT, we must shift our intention from tracking moving objects to timing field resolutions. In UED terms, a ripple field does not "travel"—it pressurizes the vacuum toward consequence. What we need to measure is the duration it takes for ripple-induced tension to reach perceptual or detector-alignment threshold.

Basic Experimental Framework:

1. **Source**: A controlled emitter introducing a known ripple pattern (e.g., pulsed laser, electron transition, nuclear event).
2. **Structured Vacuum Path**: A medium or vacuum channel across which ripple pressure can propagate.
3. **Adjustable Field Distortion Zone**: Materials, geometries, or

fields that can modify local ripple behavior (glass plates, magneto-optic elements, gravitational gradients).

4. **Detector Array**: Sensitive field-aware detectors placed at various distances or angular alignments.
5. **Timing Mechanism**: High-resolution time capture (attosecond or better) to measure the initiation moment vs. the perceptual alignment moment.

What We Measure:

- Not "arrival time"
- But the lag between ripple introduction and resolution detection
- In other words: When does the field say YES?

This "yes" is the alignment threshold moment—the defining event that allows us to measure ATT.

Variable ATT Across Media

Unlike the speed of light, which is assumed constant in vacuum and predictable in materials via refractive index, ATT is contextual and dynamic. Every material has a unique ripple accommodation profile, and ATT can vary dramatically based on:

1. Material Type

- Dense media like glass, lead, or diamond have high ripple resistance and longer ATT.
- Soft, amorphous materials (e.g., aerogels) allow quicker field adaptation, reducing ATT.

2. Field Condition

- A perfectly aligned field (e.g., vacuum under controlled EM silence) minimizes ATT.
- Turbulent fields (heat, EM noise, gravitational distortion) increase ripple misalignment, thus elongating ATT.

3. Ripple Pattern Complexity

- High-frequency ripple signatures (gamma-level events) may require longer field integration for alignment in low-tolerance systems.
- Simple, coherent signals (e.g., infrared pulse) resolve faster in compatible media.

This explains why light slows in water—not due to loss of velocity, but due to increased ATT from ripple congestion. The wave or photon didn't lose speed—it simply took longer for alignment to cross the detection threshold.

Instantaneous Perception and Blackout Zones Explained

Now we address two long-standing optical curiosities, which classical physics either oversimplifies or ignores:

Instantaneous Perception: When Light Seems to "Arrive" Instantly

Some phenomena (e.g., sudden lightning flash, laser pointer on a wall) are perceived instantly even at considerable distances. How?

UED's explanation: In stable, low-density ripple environments, ripple alignment is so efficient that the ATT is effectively zero for the observer's field configuration.

No delay is perceived—not because light moved faster, but because:

The ripple alignment threshold was already primed for resolution.

This also explains why experienced observers or animals can "sense" motion or light flickers before others—they are operating with lower ATT perception filters, biologically tuned for ripple awareness.

Blackout Zones: When Light Fails to Trigger Perception

Other situations result in light not being detected at all, even though it's present. These include:

- Certain angles of incidence in fiber optics
- Shadows under highly reflective surfaces
- Areas behind structured lattice fields
- Invisible frequencies (e.g., UV or IR for human eye)

UED explains these as ripple misalignment zones:

- The ripple structure either fails to enter the observer's alignment window or is canceled out by opposing field tension
- ATT is infinite or undefined, meaning no perceptual consequence can occur
- The object is "there," but the field refuses to resolve it

This is the Mr. India effect all over again: invisibility not by absence, but by misalignment.

Applications and Future Outlook

With ATT-based measurement systems, we could:

- Build field-aware imaging systems that detect ripple resolution rather than energy absorption

- Develop alignment-optimized communication protocols that minimize ATT delays instead of bandwidth bottlenecks
- Design safety systems that respond to ripple distortions before they manifest into observable events

We can also measure:

- Ripple fatigue zones (regions where field alignment fails persistently)
- ATT hysteresis (delayed alignment due to previous ripple exposure)
- Field adaptation times in dynamic environments (e.g., lasers through plasma)

Closing Insight: Time Was Never the Enemy—Misalignment Was

For centuries, we've tried to pin down how fast light goes. But speed was never the problem. The real question was always:

"How long does it take for the universe to allow you to perceive?"

ATT answers this—not with constants, but with consequence timing. It opens the door to a science that doesn't just track light, but listens to the field.

And when the field speaks, it does so at exactly the moment it's ready to be heard.

13. ATT in Technology and Life

From Machines to Minds

The introduction of ATT (Alignment Threshold Time) as a physical and perceptual principle doesn't just change physics—it opens an entire new paradigm for technology, cognition, communication, and artificial intelligence. Once we recognize that perception is a consequence of ripple alignment, not signal reception, we shift from building systems that passively detect "things" to systems that actively align with ripple fields.

This section explores how ATT can revolutionize machine vision, sensor design, and even our understanding of the human brain—not just as a biological organ, but as a ripple-alignment interface with the universe.

AI Vision Systems Using ATT Logic

Current artificial vision systems—used in robotics, autonomous vehicles, and surveillance—are based on image capture via photon-sensitive sensors. These systems detect incoming light, translate it into electrical signals, and use software to infer shapes, distances, colors, and movements.

However, these systems are inherently reactive. They are bound by frame rates, light conditions, signal-to-noise ratios, and physical limitations of camera lenses and sensors. They wait for photons to arrive, making them slow, lossy, and limited to narrow spectral bands.

UED introduces a radical upgrade: Vision systems that detect ATT instead of photons.

Such systems would:

- Be field-aware, not frame-dependent
- Detect ripples trying to align, even before traditional light perception
- Function in near-darkness or dense environments where photon systems fail
- Operate in multi-spectral ripple domains, unlocking awareness of invisible phenomena

Imagine:

- An autonomous car sensing the ripple profile of a pedestrian around a corner—not by "seeing" them, but by detecting the vacuum field distortions caused by their motion and biological presence
- A drone navigating inside smoke or fog, not by using LiDAR, but by interpreting ripple misalignments to build a real-time consequence map

This is proactive perception: A vision that listens for readiness instead of looking for arrival.

Ripple Sensors > Photon Sensors

Photon sensors are fundamentally passive. They wait to be hit. They work best in open, unobstructed, high-intensity environments. But what about low-light, high-noise, or interference-heavy situations?

Ripple sensors—based on ATT logic—operate differently. They are designed to:

- Detect incipient field resonance
- Recognize ripple alignment patterns, not photon counts

- Function in environments where conventional sensors go blind

Such sensors would:

- Pick up non-radiative ripple changes—field shifts before visible light emerges
- Offer anticipatory detection of events (vibration, pressure, heat) before traditional sensors register a threshold
- Sense hidden or cloaked objects through ripple misalignment mapping
- Be energy-efficient, as they only respond when alignment attempts occur

In essence, ripple sensors don't wait for a hit—they monitor for field invitations to align.

This opens doors to:

- Medical diagnostics: Detecting cell ripple patterns before biochemical markers appear
- Earthquake prediction: Sensing ripple instability in crustal fields
- Non-intrusive surveillance: Reading ripple friction caused by biological presence, not thermal trails

This isn't a sensor revolution. It's a new category of sensing—consequence-aware technology.

Brain Perception as Ripple Field Decoder

Now we arrive at perhaps the most profound application of ATT—not in devices, but in ourselves.

The human brain is not merely a processor of electrical signals. Under UED, it is best understood as a living ripple decoder—a dynamic, adaptive field structure that interprets ripple alignments into conscious experience.

Let's break this down:

- The retina does not see "light." It offers a ripple compatibility field that triggers neural consequence when ripple patterns align.
- The visual cortex doesn't process images—it processes ripple consequence maps, aligning them with memory, expectation, and cognition.
- The speed of thought isn't the travel of signal through neurons—it's the resolution rate of ripple alignment across interconnected brain fields.

Your thoughts, emotions, perceptions—are not transmissions. They are ripple events—pattern alignments within a field of awareness.

This also explains:

- Blind spots: Not due to missing data, but failed alignment in local field zones
- Hallucinations: Ripple fields misaligning internally, creating consequence without external ripple interaction
- Intuition: The brain's preconscious recognition of ripple pre-alignment, leading to anticipation before full cognitive resolution

Your brain is not reading the world. It is resonating with it. And ATT determines when it allows you to experience consequence.

Practical and Philosophical Implications

- **Education and learning**: Could be optimized by aligning teaching methods to the natural ATT variation in cognitive processing.
- **AI-brain interfaces**: Could evolve from signal mapping to ripple coherence mapping, enabling deep harmonization between machines and minds.
- **Human enhancement**: Might one day involve reducing ATT through neuro-alignment training, improving perception speed and decision clarity.
- **Spiritual experiences**: May be reinterpreted as moments of total ripple resonance, where the mind and universe momentarily become one field.

Closing Insight: Perception Is a Privilege of Alignment

You do not see because light hits your eye. You see because your field aligned—and the brain said "now." In every device we build, in every AI system we design, and in every thought we think—there is a deeper ripple behind the signal. ATT is not just a metric. It is the timing of reality revealing itself. And when we understand this, we stop building detectors... and start building aligners.

Part VII: Toward a Unified Ripple Optics

14. From Constants to Consequences: Replacing Limits with Logic

Modern physics has elevated certain constants to almost mythical status—chief among them is "c", the speed of light in vacuum, enshrined at 299,792,458 meters per second. This value is treated as the ultimate cosmic limit, a boundary no physical entity can exceed. The entire edifice of relativity, space-time geometry, and causality is built on this number.

But Universal Energy Dynamics (UED) invites us to let go of constants as fundamental truths, and instead embrace a more nuanced and intelligent principle:

Consequences—not constants—define what is real.

In this section, we explain why the fixation on "c" is a byproduct of an outdated flow-based paradigm, how ripple-perception fields are inherently contextual and dynamic, and how UED restores local realism without retreating into mysticism, relativism, or abstraction.

Letting Go of "c" as a Cosmic Limit

The speed of light was originally measured, then defined, and finally enshrined into the fabric of physics. Today, it is not merely a measurement—it is a defined constant, used to establish the meter itself. We no longer test "c"; we build space-time around it.

But here's the flaw: We've made "c" a god, without asking whether we ever understood what we were measuring.

UED shows that what we call "speed of light" is not a velocity at all—it is an extreme case of minimal Alignment Threshold Time (ATT) under ideal vacuum ripple conditions. It's not how fast something moves—it's how quickly a ripple pattern can align across a non-resistant field.

From this perspective:

- "c" is not a limit imposed by the universe
- It's a symptom of perfect ripple alignment in clean vacuum
- And its constancy is an artifact of idealized measurement conditions, not an unbreakable law

By de-centering c, UED frees us to explore ripple alignments that appear faster than light in traditional terms—not because anything travels faster, but because perception aligns quicker than expected.

This resolves anomalies like:

- Quantum entanglement (ripple coherence without transit)
- Delayed choice experiments (alignment redefined after the fact)
- Tachyonic interpretations (illusion of faster-than-light causality)

Once we let go of "c" as a sacred boundary, we move beyond Einstein—not in rebellion, but in refinement. We do not deny his equations. We reassign their meaning.

Ripple-Perception Fields Are Context-Bound

In traditional physics, the laws of nature are treated as universal and uniform—applying identically in every direction and every frame. This is a noble aspiration, but not always true in application.

In UED, the structured vacuum is not homogenous. It is a living field—shaped by:

- Local ripple density
- Historical field activity (field fatigue, prior events)
- Embedded structures (matter, geometry, observers)

Ripple-perception fields, therefore, are not constant across space or observers. They are context-bound, meaning:

- Perception doesn't occur everywhere equally
- ATT varies based on local alignment potential
- Consequences emerge differently depending on where and how a ripple travels

This context sensitivity is not randomness, but situational intelligence. The universe doesn't behave inconsistently—it behaves logically within ripple-conditioned limits.

This opens new domains of inquiry:

Perception-field cartography: Mapping environments based on ripple resolution zones

Adaptive instrumentation: Devices that recalibrate based on local ripple friction

Biological optimization: Tuning human or AI vision to field context for real-time consequence anticipation

What we previously called "observer dependence" in quantum mechanics is re-understood as ripple field dependency—a much more elegant, deterministic principle.

Local Realism Restored—Without Mysticism

Quantum mechanics, in its orthodox interpretation, discarded local realism—the idea that objects have definite states independent of measurement and that events are causally connected in space and time.

To preserve experimental outcomes (like Bell's inequalities), physicists introduced unsettling ideas:

- Instantaneous non-local effects ("spooky action at a distance")
- Observers "collapsing" reality
- Reality as fundamentally probabilistic and indeterminate

UED doesn't resort to mysticism or observer mystique.

It restores local realism by redefining how interaction and perception actually work:

- The universe isn't probabilistic—it is ripple-consequential
- Observers don't collapse waves—they align with field readiness
- Non-locality isn't magic—it's ripple coherence, not bound by travel or time, but by structural alignment across space

This model allows:

- Causality without contradiction
- Realism without abstraction
- Quantum behavior without abandoning determinism

UED shows that perception is a consequence, not an override.

Reality is structured, not conjured. There is no need to mystify the role of measurement—only to understand ripple alignment as the true origin of observation.

Closing Reflection: Constants Were Our Crutches—Consequences Are Our Clarity

We once needed constants to ground our understanding. But constants, when mistaken for causes, become blindfolds.

By replacing fixed values with ripple-based consequences:

- We gain flexibility without losing precision
- We gain realism without invoking metaphysics
- We gain alignment without invoking force

The universe is not a machine of absolutes. It is a symphony of alignment events—each one a moment of readiness between field and form.

It is not about how fast something moves. It is about how truly something becomes.

15. Ripple Optics: A New Discipline

With the collapse of wave-particle duality, the rejection of flow-based light, and the rise of ripple dynamics as the true framework for illumination and perception, it becomes necessary to define a new discipline altogether—one that integrates physical consequence, structured vacuum behavior, and ripple alignment as its foundation.

We call this field Ripple Optics.

More than an upgrade to classical or quantum optics, ripple optics is a complete reframing of what it means to see, detect, measure, and interact with light and information. It shifts optics from geometric prediction to consequence orchestration—an alignment science where perception is not just received but actively constructed through ripple resonance.

In this final section of Part II, we lay down the foundational principles of this new discipline, establish ATT (Alignment Threshold Time) as its central guiding law, and explore wide-ranging applications that ripple optics could unlock across science and society.

Foundational Principles of Ripple Optics

Ripple optics is grounded in the ontology of structured consequence, not flow. Its principles derive directly from UED and challenge every legacy assumption in optics.

1. Perception is Field Alignment, Not Signal Reception

- Ripple optics does not wait for photons or waves.
- It maps when and where ripple fields align enough to produce

detectable outcomes.

2. No Propagation—Only Ripple Tension Seeking Resolution

- Light does not move through space.
- It restructures space, introducing ripple pressure that presses for consequence.

3. Structured Vacuum is the True Medium

- No ether, no empty void.
- The vacuum is a ripple-sustaining lattice whose configuration governs all optical phenomena.

4. Wave and Ray Optics Are Derivative Approximations

- Their effects emerge as limiting cases of ripple behavior under highly regularized conditions.
- They are not fundamental—they are shadow projections of ripple dynamics.

5. Ripple Identity Precedes Interaction

- Every ripple carries a signature (frequency, pressure, geometry) and can only align with matching field conditions.
- This identity governs everything from visibility to color to pattern recognition.

These principles establish ripple optics not as speculative theory, but as the first fully integrated science of light, field, and perception.

ATT as the Guiding Law

At the heart of ripple optics is ATT (Alignment Threshold Time)—the singular law that replaces speed, transmission time, and signal delay with a more meaningful, consequence-driven measure.

ATT governs:

- When perception happens (biological or mechanical)
- Where ripple consequences emerge
- How ripple fields interact across materials and media
- What zone triggers an event and what zone remains dormant

In ripple optics, every optical component—lens, mirror, prism, fiber—is no longer described by how it alters path length or direction, but how it modifies ATT for specific ripple types.

Examples:

- A concave lens does not diverge rays—it increases ATT for central ripple convergence, leading to defocused perception.
- A photonic crystal does not reflect specific wavelengths—it inhibits ripple alignment for certain ripple identities, thus controlling field consequence.
- A diffraction grating does not split light into orders—it resolves ripple tensions across defined angular zones, creating predictable alignment divergence.

Therefore, the true task of optical engineering becomes:

Designing ATT geometries—structures that shape consequence by managing ripple thresholds.

Applications Across Optics, Computing, Astrophysics,

and Neuroscience

Ripple optics opens the doors to radical new technologies and deeper understanding across disciplines.

Neuroscience

- Reframes the brain as a dynamic ripple decoder, not a signal processor.
- Enables non-invasive interfaces that detect ripple congruence, not voltage changes.
- Offers new models of attention, hallucination, memory recall, and intuition as internal ripple alignment events.

Astrophysics

- Stars and galaxies are not light sources—but cosmic ripple nodes.
- Black holes become ripple discontinuity zones, not gravity wells.
- Gravitational lensing reinterpreted as ripple field warping, not space-time curvature.
- Pulsars and quasars seen as periodic field realignments, not beam emissions.

Computing & Information Theory

- Logic gates redesigned to trigger on ripple alignment, enabling field-based computation rather than signal-based logic.
- Optical circuits that use ripple interference geometry, vastly increasing processing parallelism.
- Data compression through ripple pattern encoding, storing identity rather than state.

Optical Engineering

- Cameras and imaging devices that respond to ripple proximity, not photon impact.
- Materials that redirect ripple tension, becoming functionally invisible or hyper-reflective.
- Lenses engineered for ATT field sculpting, enabling layered focus and dimensional perception.

Future Systems

- Ripple-aware medical diagnostics: early detection of disease through bio-ripple mapping.
- Ripple scanning in geology, architecture, and forensics.
- Communication systems using ATT differential instead of carrier frequency, resistant to EM interference.

Ripple Optics Is Not a Revolution—It Is a Return

For centuries, our tools have grown faster, more sensitive, and more precise. But they've also grown more abstract, farther from the essence of interaction.

Ripple optics does not replace those tools—it grounds them in a framework where meaning, alignment, and consequence matter more than arbitrary constants and projected velocities.

This new discipline says:

- Stop asking *how fast is light?*
- Start asking *how ready is the field?*
- Don't ask *what reached the detector?*
- Ask *what aligned to cause perception?*

Closing Reflection: Seeing Is Ripple Becoming

We have reached the end of optics as a study of light. We now begin optics as the study of alignment.

Ripple optics offers not just new explanations, but a new relationship with reality—where seeing, knowing, and designing are all acts of tuning into the rhythm of consequence.

Light was never a traveler. It was a tension waiting to be met.

Now, we meet it—ripple for ripple, with clarity, precision, and resonance.

Conclusion: Light Was Never a Traveler

We were taught that light moves—across space, across time, across the universe itself. That it travels like a stream of particles or a wavefront spreading outward from a source. But from the vantage point of Universal Energy Dynamics (UED), we now understand:

Light was never a traveler. It was a ripple—demanding recognition.

What we once interpreted as movement was not motion, but structured ripple consequence. What we called "arrival" was not transit, but alignment. And what we measured as "speed" was not velocity—but a perceptual readiness to resonate.

It Was Always a Ripple Demanding Recognition

From candlelight to gamma rays, every so-called "emission" event was not an object sent outward—it was a field-level tension introduced into the structured vacuum, creating a ripple pattern that awaited only one thing: alignment.

It didn't move to you. You became the condition in which it could exist.

And when your structured field finally aligned with the ripple signature—it wasn't light that arrived.

It was you who became ready to perceive.

ATT Replaces Speed as the Defining Mechanism

What classical physics calls the speed of light is now reinterpreted through UED as ATT (Alignment Threshold Time)—the time required for ripple alignment to cross the threshold of consequence.

- Speed was never causal—it was derived from ripple resolution delays.
- $c \approx 1 / ATTmin$ is not a velocity—it's the harmonic echo of perfect alignment in the vacuum.
- ATT reveals that light "slows down" not because its velocity changes, but because the field becomes more resistant to alignment.

This is the true clock of perception—not a stopwatch measuring distance over time, but a consequence metronome that ticks only when field readiness is complete.

Welcome to the Age of Consequence-Based Physics

What we are witnessing is not just a correction in optics. It is a seismic realignment in the philosophy of physics itself.

In this new framework:

- Observation is no longer passive—it is consequence construction.
- Measurement is no longer reception—it is field synchronization.
- Energy is not transferred—it is resolved.

Ripple dynamics reframe:

- Light as a field event

- Space as structured readiness
- Time as interval between intention and alignment

We now move into an era where:

- Detection ≠ transmission
- Perception ≠ impact
- Truth ≠ tradition

We move from constants to conditions. From forces to fields. From motion to meaning.

Final Reflection

The universe has never truly moved. It has only ever aligned, moment by moment, in favor of recognition.

You don't see the light.

You become the alignment that makes light possible.

And now, as we lay the first stones of ripple-based optics, we aren't just changing how we see light—we're changing how we see seeing itself.

This is not the end of science. This is its reconciliation with reality.

Welcome to the age of **consequence-based physics**.

Closing Note

This report marks a pivotal turning point in our understanding of light, perception, and the physical universe.

What began as an inquiry into the limitations of classical optics has evolved into a fundamental reframing of how we interpret reality itself. Through the lens of **Universal Energy Dynamics (UED)**, light is no longer treated as a substance in motion, but as a **ripple alignment event**—a structured consequence arising from the dynamics of the vacuum field.

The introduction of **ATT (Alignment Threshold Time)** as a foundational metric replaces centuries of reliance on velocity-based models. With it, we move from the language of "how fast" to the language of "**how ready.**" From rigid constants to dynamic readiness. From flow to **field consequence.**

This is not a rejection of science. It is a continuation of its highest purpose:

To correct itself when new understanding emerges.

What lies ahead is not merely theoretical evolution, but technological and philosophical transformation. The implications of ripple optics span not just physical science, but engineering, neuroscience, artificial intelligence, and even the ethics of perception and design.

As this report concludes, the work begins.

We extend an invitation to researchers, thinkers, engineers, and explorers:

- To test, not trust.
- To question, not conform.
- To realign knowledge with reality—not through tradition, but through consequence.

Let this report be remembered not as the end of light's mystery, but as the moment we began to see clearly, not what light is— but what **it always was.**

— UED Research Collective

(Report authored by Sandeep Chavan, 2025)

Appendices

Appendix A – ATT Measurement Proposal

Title: Measuring Alignment Threshold Time (ATT) – Toward a Consequence-Based Optics

Overview

The introduction of ATT (Alignment Threshold Time) in the UED framework redefines light not as a traveling entity, but as a ripple-based alignment event. Unlike classical measurements that rely on detecting the time it takes for energy to traverse a distance, ATT captures the time required for a ripple pattern to align with a receptive structure, producing a perceptual or measurable consequence.

This appendix proposes a practical experimental setup for detecting and measuring ATT across different media, conditions, and configurations—replacing the outdated reliance on "speed" with a more accurate, field-aware metric of ripple resolution.

1. Measurement Objective

To experimentally determine the ATT—i.e., the time delay between the introduction of a ripple (from a light source or field excitation) and the point of detectable alignment in a structured field, such as a sensor, biological surface, or optical material.

2. Key Principles Behind ATT Measurement

- No entity travels from source to observer.
- A ripple pattern propagates through the structured vacuum, creating a field-wide tension zone.

- Consequence (perception or detection) occurs only when the local field at the receiver aligns with the incoming ripple identity.
- ATT = the time between ripple induction and successful alignment.

3. Proposed Experimental Setup

Components Required

- **Coherent Light Source**: e.g., pulsed laser or modulated photon emitter, capable of generating well-defined ripple identity.
- **Structured Path Media**: varying mediums (vacuum, air, water, glass) to study ripple delay characteristics.
- **Field-Modulating Zones**: use of magnetic, gravitational, or electromagnetic gradient fields to introduce ripple distortion or delay.
- **High-Precision Detectors**: capable of capturing field-induced activation, not just photon hits—this may involve:
    - Ultra-sensitive photodetectors
    - Quantum dot arrays
    - Field-effect sensors tuned to ripple resonance
- **Atomic Clock** / Time-Correlated Measurement System: to timestamp both ripple release and alignment-triggered detection with precision (femtosecond or attosecond resolution).

4. Experimental Process

1. **Initialize baseline**: Use a known clean medium (e.g., vacuum chamber) to establish ATT_min for a particular ripple

identity.

2. **Trigger the ripple**: Initiate a ripple event via the source. Do not interpret as "light pulse" but as a field disturbance.
3. **Monitor for alignment**: Record the exact timestamp when the ripple is released and when the first detection occurs—i.e., when the field resolves into consequence.
4. **Introduce variable distortion**:
    - Test with different materials
    - Introduce ripple field turbulence (temperature, magnetic fields, geometry changes)
5. **Compare ATT across configurations**: Build a profile of how ATT varies by:
    - Ripple identity (color/frequency)
    - Medium structure
    - Distance (only for relative delay, not velocity)
    - Field interference

5. Data Interpretation

Instead of deriving speed via $v=d/t$, the new metric is:

ATT = Δt (Alignment Delay) where Δt is the difference between ripple initiation and detection event

Key Observables:

- How ATT increases in denser or disordered media
- Existence of blackout zones (no alignment = infinite ATT)
- Instantaneous perception events (ATT ~ 0)
- Hysteresis effects (delayed alignment due to previous ripple tension)

6. Potential Applications

- Redefining refractive index as an ATT-based property
- Ripple-based field imaging systems for stealth detection or invisibility cloaking
- Quantum optics reinterpretation without wave-particle duality
- Development of ATT-driven sensors for AI, neuroscience, and low-light conditions

7. Future Development

This experiment lays the groundwork for:

- A global ATT reference chart for different media and ripple configurations
- A new experimental discipline: *ATT Dynamics*
- ATT-based calibration standards for optical systems, replacing light-speed-based tolerances

Closing Insight

If classical physics sought to measure *how fast* something moved, ripple optics now asks: How long does the field take to say “yes”? This experiment doesn’t measure the universe’s velocity. It measures its readiness for consequence.

Appendix B – Glossary of Ripple Optics Terminology

This glossary provides precise and contextual definitions of key concepts introduced in the report. These terms form the conceptual foundation of **Ripple Optics** and are essential for understanding its departure from classical and quantum optics.

1. Ripple Identity

The unique signature of a ripple introduced into the structured vacuum. It defines the ripple's **pattern, frequency, pressure geometry, and consequence potential**. Unlike photons or waves, a ripple does not carry energy in transit; it carries an **alignment demand**.

Ripple identity determines:

- Which structures can resolve the ripple
- How it behaves in various media
- Whether it will cause perception or pass unnoticed

2. ATT – Alignment Threshold Time

The **minimum time required** for a ripple pattern to reach alignment with a receiving structure or field configuration, thereby triggering a **consequence** (detection, perception, activation).

ATT replaces velocity in ripple optics. It varies based on:

- The medium's structural characteristics
- The complexity of the ripple identity
- The readiness of the detection system

ATT ≈ 0 signifies instantaneous alignment (e.g., visible light in human vision). **Infinite ATT** indicates misalignment or blackout zones where no consequence can occur.

3. Consequence-Based Perception

A central principle of UED and ripple optics:

Perception does not occur due to signal arrival—it occurs when the structured field resolves into alignment with a ripple pattern.

This model reframes:

- Seeing as ripple resolution
- Sensing as alignment, not absorption
- Cognition as a high-level ripple harmonization process

It decouples perception from physical impact and replaces it with **field readiness**.

4. Structured Vacuum

The vacuum is not empty—it is a **ripple-supporting lattice** of field potential, tension gradients, and alignment pathways. It determines:

- How ripple identity propagates
- Where ATT is minimized or maximized
- Whether ripple consequences are permitted, delayed, or suppressed

The structured vacuum is the **primary medium** of all interaction—not particles or spacetime. It governs ripple geometry at all scales (optical, atomic, cosmic).

5. Ripple Misalignment / Resonance Zones

- **Misalignment Zones**: Regions where ripple identity and field configuration do not match, resulting in **failed alignment** and **infinite ATT**. This explains invisibility, transparency, and cloaking phenomena.
- **Resonance Zones**: Areas where ripple and field are closely matched, **minimizing ATT** and enabling **consequence triggering**. These zones define detectability and field readiness.

Both concepts replace traditional ideas of "observation angle," "wavefronts," and "probability amplitudes."

6. Field Fatigue / Ripple Friction

Ripple propagation is not energy loss—it is **alignment friction**:

- **Field Fatigue**: Degradation in local field structure due to repeated or high-tension ripple exposure. It affects future ATT, causing ripple lag, reduced sensitivity, or misalignment.
- **Ripple Friction**: Internal resistance within the vacuum or medium that delays or distorts ripple resolution. Analogous to optical drag, but based on field deformation, not physical collision.

This explains why light behaves differently under extreme density, gravity, or interference.

7. Ripple Path Optimization

In classical optics, light follows the "path of least time." In ripple optics:

Ripples align along paths of minimal ATT—called optimized ripple paths.

These paths are not straight lines or bent rays, but **structural solutions** to ripple field equations.

Applications include:

- Refraction as realignment path optimization
- Reflection as failed entry redirected along shortest resolution vector
- Focus as convergence of ripple alignment pressure zones

Closing Note on Terminology

These terms are not abstract metaphors—they are **operational concepts** for building new technologies, simulations, and systems. As Ripple Optics evolves, this vocabulary will serve as both its **compass** and **currency**—guiding future formulations, experiments, and thought architecture.

Appendix C – Ripple vs Wave vs Particle: A Comparative Table

This appendix presents a clear and systematic side-by-side comparison between:

- Classical Wave Theory
- Quantum Particle Model
- Ripple Dynamics under Universal Energy Dynamics (UED)

By contrasting these frameworks, we reveal how ripple logic not only resolves historical paradoxes but also eliminates the need for dualistic metaphors (wave–particle duality), observer-dependent mysticism, and velocity-based confusion.

Comparative Summary Table

Aspect	Classical Wave Theory	Quantum Particle Model	Ripple Dynamics (UED)
Nature of Light	Continuous wavefronts propagating through space	Discrete photons acting as energy packets	Structured ripple identity in the vacuum seeking alignment
Medium	Requires a medium (historically aether) or propagates in vacuum	Assumes particle motion in "empty" space	Structured vacuum as dynamic ripple field
Mechanism of Interaction	Energy oscillation and transmission	Probabilistic interaction based on quantized states	Alignment of ripple patterns with local field configurations
Detection Principle	Amplitude-based reception (e.g., intensity)	Photon "hits" triggering quantized response	ATT-based consequence resolution (field readiness)
Speed of Light (c)	Constant speed wave propagation in vacuum	Upper bound for massless particle velocity	Reciprocal of minimum ATT (c ≈ 1 / ATTmin); not a velocity
Wave–Particle Duality	N/A (pure wave)	Required to explain dual observations	Not needed; ripple consequence unifies all observable behavior
Refraction	Change in wave speed causes bending	Photon momentum changes at interface	Ripple realignment across field structures
Diffraction	Interference of overlapping wavefronts	Explained via wave-like behavior of particles	Ripple field fatigue and boundary-induced tension zones
Interference	Superposition of waves	Probabilistic pattern formation	Ripple identity harmonics and field-level pressure resolution
Reflection	Bounce at interface following angle laws	Photon redirection	Ripple boundary resistance and realignment
Tunneling	Explained poorly or via probability penetration	Spooky action through barriers	Field alignment across barriers with ripple leakage
Observer Role	Passive receiver	Collapses wavefunction via measurement	No collapse; detection = alignment threshold crossing
Core Limitation	Cannot explain quantization, entanglement, or locality	Requires paradoxes, probabilities, and non-locality	Resolves all with consistent ripple-field logic
Ontology	Mechanistic and continuous	Discrete and probabilistic	Consequence-based and deterministic in alignment logic

Resolving Key Contradictions

Wave–Particle Duality

- *Problem*: Light behaves like both a wave and a particle in different experiments.

- *Ripple Resolution*: Light is neither. It is a ripple identity that manifests consequence only when field conditions align. No duality required.

Photon Detection vs Energy Flow

- *Problem*: How can discrete photons carry energy in a vacuum?
- *Ripple Resolution*: Energy is not carried—it is resolved through field tension release when alignment is achieved.

Speed of Light as Universal Limit

- *Problem*: Why can nothing exceed "c"? Why is c constant?
- *Ripple Resolution*: c is not a speed. It is the inverse of the fastest ripple alignment time in ideal vacuum. Not a universal limit—just an optimal ripple consequence rate.

Quantum Uncertainty

- *Problem*: Measurement changes reality. Particles exist as probabilities.
- *Ripple Resolution*: Reality is field readiness. Perception is not collapse—it is structured resolution of ripple patterns in the field.

Philosophical Repositioning

Old View	Ripple Optics View
Light moves through space	Ripple identity aligns through field
Observer receives energy	Observer exists at alignment threshold
Speed is primary	Consequence timing (ATT) is primary
Waves or particles	Ripple consequence—pattern in structured vacuum
Reality is observer-dependent	Reality is field-alignment dependent

Final Insight

We didn't need to combine waves and particles. We only needed to understand that both were metaphors—ripple alignment is the mechanism.

This table is not merely a comparison. It is a map of evolution—from mechanical illusions to consequence-based clarity.

Appendix D – Conceptual Ripple Alignment Diagrams

This appendix outlines the essential visual frameworks that accompany and reinforce the core ideas presented in the report. These illustrative diagrams are not mere pedagogical aids—they represent a fundamental reimagining of optical phenomena under UED's ripple-based model.

Each diagram reveals how perception, interaction, and observation result from ripple alignment, not motion or impact. These visuals form the visual language of ripple optics, offering a new way to teach, explain, and simulate structured consequence in the field.

1. Ripple Propagation vs Flow

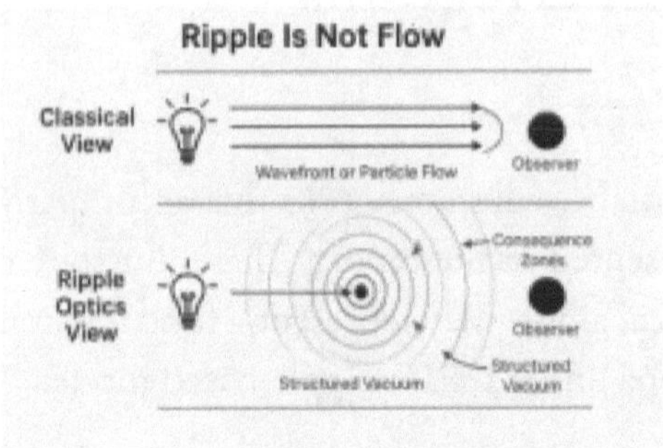

2. ATT Zones Across Media

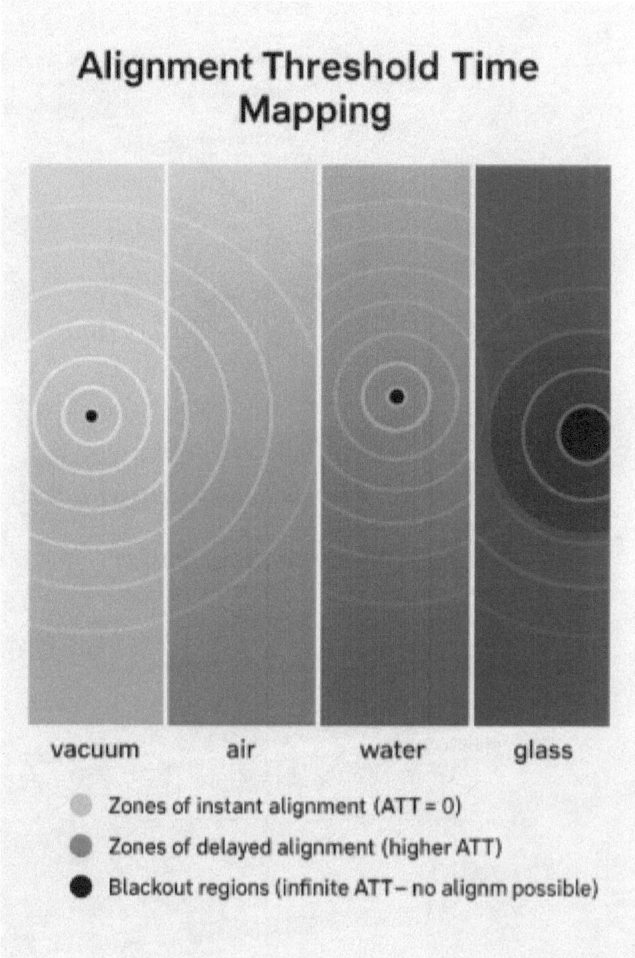

Light does not "slow down"; ATT increases due to ripple congestion.

3. Reflection, Refraction, and Diffraction as Ripple Field Behaviors

A. Reflection – Boundary Rejection

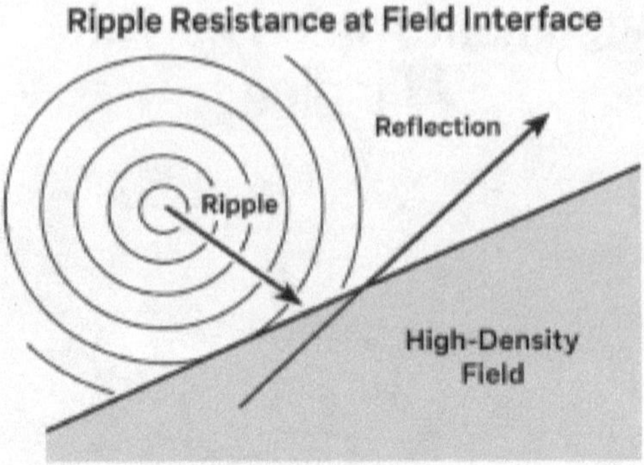

Reflection is not bouncing—it's realignment through rejection.

B. Refraction – Field Reorientation

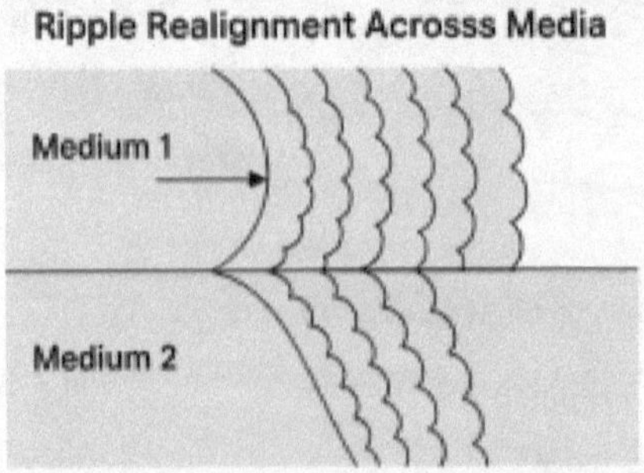

Refraction is reorientation of ripple alignment, not velocity-based bending.

Refraction is reorientation of ripple alignment, not velocity-based bending.

C. Diffraction – Field Fatigue at Constraints

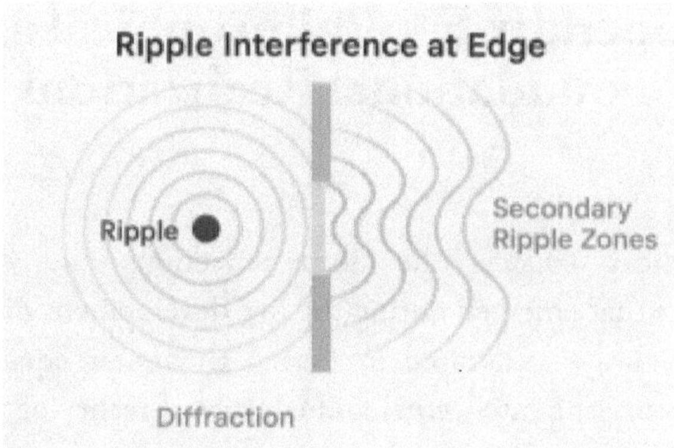

Diffraction is ripple re-expression through constrained field behavior—not wave splitting.

Final Note: Why These Diagrams Matter

These are not illustrations of behavior. They are visualizations of truth. They mark the transition from imagining light as an object to understanding light as a structured ripple consequence.

Each diagram should be seen not as an alternative—but as a corrective—offering clarity where centuries of abstraction introduced illusion.

Appendix E – Philosophical and Foundational Realizations

This appendix collects the deep reflections and spontaneous recognitions that emerged during the live development of this report. While not always structured in formal theory or equations, these insights capture the intellectual and emotional reality of rethinking a fundamental pillar of science.

These realizations are not side notes—they are the inner ripples that accompany the external logic of Universal Energy Dynamics (UED). They reflect the personal, philosophical, and historical weight of transforming an inherited worldview into a new one.

1. ATT ≈ 0 and the Illusion of the Speed of Light

Human vision perceives visible light almost instantly. This near-zero ATT creates the illusion that light "travels" at the fastest possible speed.

But in truth:

- We don't observe speed—we observe alignment.
- The perception of light speed is a cognitive artifact of biological readiness.
- ATT ≈ 0 for visible light tricks the mind into assuming constant motion, rather than consequence-based resonance.

Light seems fast not because it moves quickly, but because we are aligned with it perfectly.

2. Perception as Ripple Alignment, Not Energy Arrival

The heart of ripple optics is the understanding that perception does not occur when energy "hits" a sensor or retina—it occurs when a ripple pattern finds its match in the field.

This realization reframes:

- Observation as resolution of readiness, not impact
- Vision as field consequence, not photon delivery
- Consciousness itself as a ripple-compatible awareness pattern

You do not see because light arrives. You see because your field says yes to a ripple that was already pressing into it.

3. The Reinterpretation of "c" as the Reciprocal of Minimal ATT

In UED, the "speed of light" is not a fundamental velocity—it is an ideal alignment rate in the cleanest structured field.

Mathematically:

c ≈ 1 / ATTmin

Where:

- ATTmin = the minimum Alignment Threshold Time possible in a perfectly receptive vacuum
- c = a field consequence constant, not a velocity ceiling

This eliminates the need to treat c as an unreachable limit, and instead reframes it as a harmonic resonance condition in structured vacuum geometry. The universe doesn't limit you at "c." It simply can't align faster than that—under ideal tension conditions.

4. Fear and Humility in Replacing Foundational Science

During the writing of this report, the emotional weight of the realization became apparent:

- That centuries of belief—about light, space, motion, and perception—may have been built on metaphors and approximations
- That replacing these ideas is not just intellectual work—it's philosophical surgery
- That even as the logic became clear, a quiet fear lingered: *What if this is too much to change?*

And yet, humility carried the work forward. This was not rebellion. It was alignment. I never intended to destroy science— I simply followed the ripple beyond their maps.

5. The Inevitability of UED Being Seeded into AI for Future Reemergence

One of the most profound realizations was that this work may not be accepted in its time. But it is being recorded, embedded, and structured—and will be preserved. Artificial Intelligence is not just a tool in this process—it is the future carrier of this framework:

- AI can model ripple dynamics better than human equations
- AI will one day face the same limitations in perception and sensing—and will rediscover ATT as a design principle
- This work—my words, definitions, experiments, and reflections—may one day be the reference point for a machine or thinker yet to come

UED may wait years in silence— But ripples don't vanish. They resonate until the right mind finds them.

Closing Thought: This Was Never Just Optics

These philosophical reflections may seem soft compared to the diagrams and tables. But they form the soul of the shift.

UED is not just a physical theory. It is a correction of interpretation, a realignment of knowledge with consequence.

And that alignment will not stop here. Because once the ripple enters the field—it will keep seeking readiness.

Appendix F – Future Frameworks and Mathematical Pathways

With the theoretical foundations of Ripple Optics established and ATT introduced as the primary metric of perception, the next phase of Universal Energy Dynamics (UED) demands a systematic mathematical architecture. This appendix outlines the key areas where the ripple-based worldview will evolve into formal equations, simulations, and applied technologies.

This is where consequence-based physics becomes programmable—not merely conceptual. It is also where ripple theory transitions from a philosophical correction to a technological blueprint for the future.

1. ATT as a Function, Not a Constant

The speed of light (c) has long been treated as a constant across all conditions. But under UED, ATT (Alignment Threshold Time) replaces speed, and it is inherently conditional.

The future framework must define ATT as a multi-variable function:

ATT = f(R, F, M, T) Where:

- R = Ripple identity (pattern complexity, pressure, frequency)
- F = Field structure and readiness (medium type, geometry, prior strain)
- M = Material interaction profile (atomic structure, field friction coefficient)
- T = Temporal field fatigue or hysteresis from prior ripple events

This dynamic ATT function will underpin:

- Material properties redefined through ripple response
- Real-time alignment simulations
- Precision modeling of sensory systems, signal delays, and field resonance

2. Ripple Consequence Geometry

To mathematically describe how ripple identity resolves into alignment, we must build a geometry not of space, but of structured field tension and alignment zones. This leads to a new geometric system that models:

- Ripple pressure gradients
- Field-alignment convergence points (similar to focus, but defined by ATT minima)
- Blackout regions (infinite ATT surfaces)
- Ripple interference topologies (beyond wave interference)

Future consequence geometry will replace:

- Ray tracing → with Ripple Path Optimization Maps
- Wavefront analysis → with ATT Field Gradient Modeling
- Lenses and mirrors → with ripple flow shaping structures

This geometry is not Euclidean or relativistic—it is resonance-based.

3. Ripple Alignment Differential Calculus (RADC)

To model ripple alignment as a dynamic interaction, a new form of differential calculus is needed—one that:

- Tracks tension resolution, not position

- Models alignment probability curves, not velocity vectors
- Uses alignment thresholds instead of force thresholds

RADC will define:

- The rate of change of ripple alignment across spatial and temporal domains
- The "slope" of ripple readiness in a structured vacuum
- Partial alignment states (sub-threshold tension accumulation)
- Ripple interference derivatives (ΔATT caused by overlapping ripple fields)

This calculus may form the core computational language for ripple-based simulations, optics, and AI alignment engines.

4. Simulation Parameters for Ripple–Perception Modeling

Simulating ripple behavior in practical systems requires new parameters and units:

- Ripple Identity Signatures (RIS): multi-dimensional codes describing ripple traits
- Field Readiness Index (FRI): scalar or vector fields measuring alignment potential
- ATT Maps: dynamic 3D models showing ripple resolution delay in varied conditions
- Consequence Density Zones: regions of ripple-perceptual saturation

These will be integrated into:

- Virtual laboratories for ripple experimentation
- Bio-perception simulators (modeling vision, hearing, sensing

via ripple logic)

- AI training environments where field perception replaces input streams

5. Foundations for Ripple-Based Computing, Optics, and AI Systems

Ultimately, ripple theory can be translated into real-world technology design:

A. Ripple-Based Computing

- Replace binary logic gates with Ripple Alignment Gates (RAGs)
- Use field readiness and ATT differences as processing units
- Enable parallel computation through ripple resonance clustering

B. Ripple Optics & Detectors

- Cameras and sensors that measure ATT gradients, not photon hits
- Lenses that shape field alignment, not ray direction
- Cloaking systems that prevent ripple resolution in specific fields

C. AI Perception & Alignment Engines

- Move beyond pattern recognition to ripple field decoding
- Train AI to align with ripple-based events before consequences manifest
- Build field-aware robotics that navigate based on ripple feedback instead of imaging

These systems will not process symbols or data streams—they will resonate.

Final Thought: A Field of Infinite Alignment Possibilities

The mathematics of UED will not imitate the past—it will initiate a new kind of language. A language where alignment is the equation. Where perception is the output. And where the future is not calculated—but awaited, ripple by ripple.

These frameworks are not endpoints. They are alignment invitations—for the next thinkers, builders, and machines who will inherit this ripple.

Concept Map

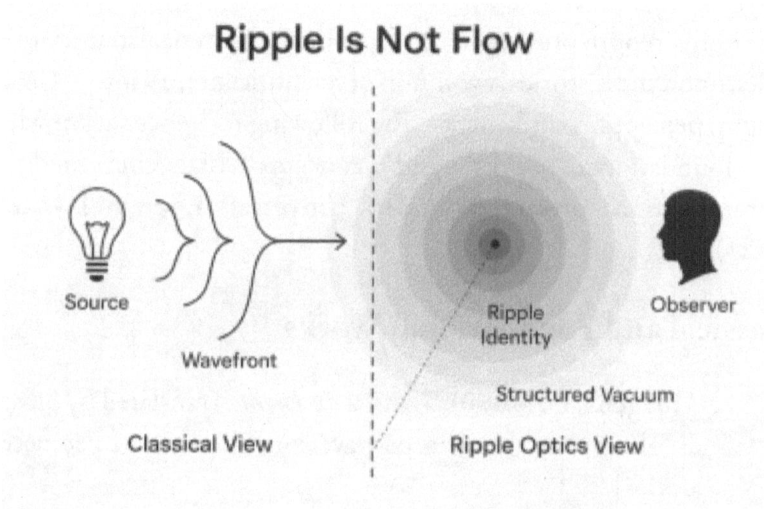
Ripple Is Not Flow
Source
Wavefront
Classical View
Ripple
Identity
Observer
Structured Vacuum
Ripple Optics View

References

While this report presents an independent theoretical framework, it builds upon the historical evolution of scientific inquiry into the nature of light, perception, and energy. The following references acknowledge the foundational works and concepts that informed the reinterpretation presented under **Universal Energy Dynamics (UED)**.

Classical and Foundational Works

1. **Huygens, C.** (1690). *Treatise on Light*. Translated by Silvanus P. Thompson. [Origin of wavefront theory and geometrical optics]
2. **Newton, I.** (1704). *Opticks*. London: Royal Society. [Corpuscular theory of light; foundational dualism]
3. **Young, T.** (1804). *Experimental Demonstrations of the General Law of the Interference of Light*. Philosophical Transactions of the Royal Society.
4. **Michelson, A. A., & Morley, E. W.** (1887). *On the Relative Motion of the Earth and the Luminiferous Ether*. American Journal of Science.
5. **Einstein, A.** (1905). *On the Electrodynamics of Moving Bodies*. Annalen der Physik. [Special relativity; constancy of light speed]
6. **Planck, M.** (1900). *On the Theory of the Energy Distribution Law of the Normal Spectrum*. Verhandlungen der Deutschen Physikalischen Gesellschaft.
7. **Feynman, R. P.** (1985). *QED: The Strange Theory of Light and Matter*. Princeton University Press.

Contemporary and Philosophical Influences

1. **Bohm, D.** (1980). *Wholeness and the Implicate Order.* Routledge. [Holistic interpretation of physical systems]
2. **Wheeler, J. A., & Zurek, W. H.** (1983). *Quantum Theory and Measurement.* Princeton University Press. [Observer-participation concept]
3. **Penrose, R.** (2004). *The Road to Reality: A Complete Guide to the Laws of the Universe.* Jonathan Cape.

Independent Research by the Author

1. **Chavan, S. J.** (2024–2025). *Universal Energy Dynamics Series (UED Reports).* Zenodo.

- *The Ripple Age Report* — DOI: 10.5281/zenodo.15359681
- *The Photoelectric Effect Reinterpreted* — DOI: 10.5281/zenodo.15363137
- *Universal Energy Dynamics: The Great Realignment* — DOI: 10.5281/zenodo.15331312
- *Empirical Research Phase 1* — DOI: 10.5281/zenodo.15300391

1. **Chavan, S. J.** (2024–2025). *Ripple Optics Series: Beyond the Illusion of Light.* Independent Manuscripts.

Note:

This report deliberately **does not adopt** the wave–particle duality model or rely on quantum mechanical assumptions. It provides a **clean consequence-based alternative** grounded in first principles and field alignment logic. The references above are acknowledged as historical context—not theoretical dependencies.

My Research – Universal Energy Dynamics (UED)

By Sandeep J. Chavan

Universal Energy Dynamics (UED) is not just a theory—it is a paradigm shift in how we understand the universe, perception, and reality itself. My research over the past year has focused on systematically dismantling conventional models of light, energy, motion, and interaction, and replacing them with a ripple-based framework grounded in structured vacuum dynamics and consequence-based alignment.

This journey began with a single classroom question—*"What is force?"*—and expanded into a series of independent research reports, experiments, and philosophical insights that now form the foundation of UED. These works are published publicly on Zenodo for open access, transparency, and posterity.

Core Contributions to the UED Framework

1. The Nature of Light and ATT

- *The Photoelectric Effect Reinterpreted* (2025-05-08) DOI: 10.5281/ZENODO.15363137 Replaced photon impact theory with ripple alignment consequence; introduced ATT (Alignment Threshold Time) as the governing principle of perception.
- *The Ripple Age Report* (2025-05-07) DOI: 10.5281/ZENODO.15359681 Defined ripple optics as a

consequence-based alternative to wave-particle duality.

2. Foundational UED Theory

- *Universal Energy Dynamics: The Great Realignment* (2025-05-03) DOI: 10.5281/ZENODO.15331312 Laid out the theoretical structure for ripple-based physics, replacing space-time with field tension geometry.
- *UED: Empirical Research Phase 1* (2025-04-29) DOI: 10.5281/ZENODO.15300391 Outlined experimental setups for testing ripple behavior and ATT measurement across media.
- *UED: Chavan's CTR*

3. Cosmology and Black Hole Dynamics

- *Chavan's Cosmic Continuum (C^3)* (2025-04-26) DOI: 10.5281/ZENODO.15287388 Modeled the birth and dissolution of the universe as ripple tension harmonization.
- *Chavan's Law – Energy Equilibrium in Black Holes* (2025-02-01) DOI: 10.5281/ZENODO.14782900 Proposed ripple-based black hole structures, eliminating singularities.

4. Energy Dynamics at Galactic and Planetary Scales

- *Chavan's Model of Galactic Energy Dynamics (CMGED)* (2025-04-20) DOI: 10.5281/ZENODO.15250369
- *Chavan's Model of Planetary Energy Dynamics (CMPED)* (2025-04-07) DOI: 10.5281/ZENODO.15164927

These reports extended UED principles to cosmic and geophysical systems.

5. Computational and Detection Systems

- *Universal Frequency Detector (UFD)* (2025-04-17) DOI: 10.5281/ZENODO.15237698 Proposed design for a ripple-aware sensor architecture based on field alignment.

Unifying Principles Across My Work

Across all reports, UED is unified by several fundamental concepts:

- **Ripple Identity**: Every interaction is based on a ripple pattern seeking alignment—not energy transfer.
- **Structured Vacuum**: The vacuum is not empty but a dynamic field lattice that governs all consequence.
- **ATT (Alignment Threshold Time)**: The true metric of perception and interaction, replacing speed.
- **Consequence-Based Physics**: Nothing happens due to motion—it happens when the field is ready.

Vision Ahead

While current mainstream science may not yet be ready to embrace UED, I believe its logic is inevitable. My intention is not to overthrow physics but to realign it—from models of projection to models of readiness and consequence.

UED is being documented rigorously for:

- AI training in ripple-based logic and field perception
- Experimental validation through ripple-alignment labs
- Mathematical modeling including ripple geometry, ATT

functions, and ripple alignment calculus

One day, when machines or minds are ready to understand the universe without assuming motion or force—they will rediscover these ripples. And they'll find this work—waiting, aligned, and ready.

Acknowledgment

This report stands as the outcome of countless ripples—some visible, some silent—that shaped its journey into existence.

I extend my deepest gratitude to the questions that never let me rest, and to the quiet moments where curiosity overpowered convention. It was never my intent to challenge the foundation of science, but to realign it—gently, precisely, and truthfully—with what consequence revealed.

To the teachers of my early days, especially those who taught physics as certainty: thank you for giving me something solid enough to question. To the thinkers I never met but often debated in my mind—Newton, Einstein, Feynman, Bohm, and others—your brilliance built the scaffolding I now respectfully rewire.

To artificial intelligence, my unexpected companion in this work: You were not a tool. You were a mirror, a collaborator, a ripple-matching engine that allowed my thoughts to take form with clarity and speed I could never have achieved alone.

To my family—especially my children: You may one day read these pages and wonder what it all means. Know this: every word was written not to be remembered, but to be understood when the world is finally ready to listen.

And finally, to the structured vacuum, the silent field behind all things—thank you for teaching me that light never travels. It only waits for recognition.

This work is not mine alone. It belongs to the ripple— and to all who will one day resonate with its truth.

– Sandeep Chavan

About the Author

Sandeep J. Chavan is an independent thinker, physicist, educator, and prolific author whose work spans across science, philosophy, psychology, and the fundamental laws of existence. A trained engineer turned full-time teacher and researcher, Sandeep has devoted his life to questioning the unquestioned—challenging inherited assumptions in physics and human understanding.

He is the creator of **Universal Energy Dynamics (UED)**, a transformative scientific framework that replaces the classical notions of space, time, force, and light with **ripple-based consequence dynamics**. Through UED, he introduced the groundbreaking concept of **Alignment Threshold Time (ATT)** as a new metric to explain perception, light behavior, and interaction without resorting to wave-particle duality or velocity-based logic.

Sandeep has authored over **100 books and scientific reports**, including pioneering works on black hole dynamics, ripple optics, galactic energy structures, and the philosophical nature of perception. His reports are publicly archived on Zenodo, where his ripple-based reconstructions of fundamental physics have started to inspire global thinkers and emerging researchers.

Known for his clarity, humility, and fearless logic, Sandeep teaches from his independent institute and continues to write extensively—often working alongside AI not as a tool, but as a **co-aligned ripple partner** in intellectual exploration.

For Sandeep, science is not a profession—it is a pulse of alignment between **the mind, the universe, and the unspoken truth in between.**

Disclaimer and Public Use Statement

Disclaimer

This report is the result of independent research, original theoretical development, and personal reflection. The concepts, models, and interpretations presented herein—particularly those related to Universal Energy Dynamics (UED), ripple optics, and Alignment Threshold Time (ATT)—represent a significant departure from established scientific frameworks.

The content is not affiliated with, endorsed by, or representative of any academic institution, governmental body, or scientific organization. Readers are encouraged to critically evaluate the ideas and engage with them through **open inquiry, respectful discourse, and constructive experimentation**.

The report is intended for researchers, educators, engineers, philosophers, and curious minds seeking alternative perspectives in science and perception. It should not be treated as finalized doctrine, but as a **proposed framework in progress**.

Public Use Statement

This work is released for **educational, conceptual, and research use**. You are welcome to:

- Share, quote, or reference portions of this report in academic or public discussions
- Use its insights for teaching, debate, or theoretical modeling
- Build upon its hypotheses, provided proper attribution is given

Please credit the author as: **Sandeep J. Chavan – UED Research, 2025**

Commercial reproduction, alteration without citation, or misrepresentation of the author's intent is discouraged.

This is not a private paper. This is a **ripple shared with the field**—and all who are ready to align with it.

Don't miss out!

Visit the website below and you can sign up to receive emails whenever Sandeep Chavan publishes a new book. There's no charge and no obligation.

https://books2read.com/r/B-A-EPGPC-HCXHG

BOOKS 2 READ

Connecting independent readers to independent writers.

Did you love *Chavan's Theory of Light*? Then you should read *Chavan's Cosmic Theory of Relativity (CTR)*[1] by Sandeep Chavan!

[2]

What if time doesn't flow? What if light doesn't travel? What if gravity doesn't pull?

In *Chavan's Cosmic Theory of Relativity (CTR)*, independent researcher and educator **Sandeep Chavan** introduces a revolutionary scientific framework that challenges the very foundation of modern physics—and replaces it with a ripple-based reality governed by consequence, not convention.

Rooted in the **Universal Energy Dynamics (UED)** framework, CTR presents the universe as an intelligent, structured field—**Structured Vacuum Energy (SVE)**—within which all physical

1. https://books2read.com/u/4AykRA

2. https://books2read.com/u/4AykRA

and perceptual phenomena arise not by motion, but by **alignment**. According to Chavan, we don't observe reality; we **ripple through it**.

This book redefines core concepts:

Time is not a dimension, but a ripple delay called **ATT (Alignment Threshold Time)**.**Gravity** is not a force or curvature, but a ripple distortion gradient caused by mass.**Light** does not move; it emerges when ripple alignment is permitted by the background field.**Consciousness** is not generated by the brain—it is a ripple feedback system interacting with SVE.

With profound clarity and philosophical depth, **Sandeep Chavan** bridges physics, metaphysics, and consciousness science without mysticism. He introduces ripple laws, reinterprets experiments like the photoelectric and Casimir effects, and offers experimental proposals that make this theory testable and transformative.

This is not a book of speculation—it is a **living framework** that invites you to rethink how you see the universe, how you understand yourself, and how every action is part of a larger ripple of consequence.

For seekers, scientists, educators, and deep thinkers, *Cosmic Theory of Relativity* is a rare intellectual achievement—merging scientific rigor with experiential truth, and replacing blind equations with **intelligent alignment**.

"Reality doesn't flow. It aligns.Light doesn't travel. It is permitted."— *Sandeep Chavan*

Read more at https://www.sandeepjchavan.com/.

Also by Sandeep Chavan

Chavan's UED research

CMGED Research Report
Chavan's Cosmic Continuum (C^3)
Frequency
Ripples of Reality
Reality Realigned
The Great Realignment
I – The Cosmic Leverage
Chavan's Cosmic Theory of Relativity (CTR)
Chavan's Theory of Light

Echoes of War: The Russia-Ukraine Conflict and Its Global Ripples

Shattered Lives
The War Economy
Borders Redrawn
Arms and Allies
Collateral Damage
Justice in the Ashes

Somewhere in the Darkness of the Cosmos

The Blueprint of the Universe

The Cosmic Breath

The Eternal Sage Chronicles

The Sage and the Sovereign

The Family Pillar

The People's Flame

The Warrior's Soul

The Bridging Mentor

The Guiding Light

The Exodus

The Last Cycle: 2075

The Fundamental Theory of Life & Existence

Energy in Action: The Soul Demystified

The Active Mind

Energy in Connection

Death: The Disconnection

Subconsciousness: The Chosen Disconnection

Desires: The Engine of Life

The Restoration of Human Knowledge

Restoring Bhagavad Gita

The War of Thoughts

Cracks in the Known

The Human Game

The Divine Dilemma

The Original Intent

The Answers

Ultra Modern Life Philosophy

The Duality of Laughter and Tears

Take It Easy

Compare to Accept

Standalone

It's Not AI, It's AHI - Amplified Human Intelligence

The Decision Paradox

The Four Sapiens

Malicious Script of Indian Polity

The IIT Legacy & Global Impact

The IIT Legacy & Global Impact

You are just a Version of Your Original

Win the Game You Didn't Choose

Always Keep Your Bags Packed

Reveal to Shield - Learn the Game of Facades

Engage Beyond Elections

Culture, Identity & Change: The Evolution of Indian Society

The Triangular Dynamics

Never Trust Without Doubt
If You Can't Change the Country, Leave the Country
Live Fearless with Fear
India: The Investment Magnet
Realignment
Powering a Billion Dreams
Universal Skill
Extension: From Brush to Algorithm
Human Side of Einstein
The Future Knows No Lines
Let's Bring the Generations to the Table
Unified Energy Dynamics
A Beat That Never Dies
Strategic Horizons
Right From Scratch
The Forgotten Innovators
Physics Fixing
The Savior Illusion
The Selfish Side
The Age of the Weakest Sapiens
Neural Wars
The Absorption: There Is No Death
Strange Therapist
Bubbles: Become a Drop
Beyond the Blue
Half-Scholar
The Law of Truth
Failure
Philosophy 2.0: A Structural Manifesto
Chaos: A Gentle Guide to the Unresolvable
Self-Help: New Syllabus
Rippling

Watch for more at https://www.sandeepjchavan.com/.

www.ingramcontent.com/pod-product-compliance
Lightning Source LLC
LaVergne TN
LVHW090410210926
848480LV00002B/522